Design and Safety Assessment of Critical Systems

Design and Safety Assessment of Critical Systems

Marco Bozzano
Adolfo Villafiorita

CRC Press
Taylor & Francis Group
Boca Raton   London   New York

CRC Press is an imprint of the
Taylor & Francis Group, an **informa** business
AN AUERBACH BOOK

Auerbach Publications
Taylor & Francis Group
6000 Broken Sound Parkway NW, Suite 300
Boca Raton, FL 33487-2742

© 2011 by Taylor and Francis Group, LLC
Auerbach Publications is an imprint of Taylor & Francis Group, an Informa business

No claim to original U.S. Government works

Printed in the United States of America on acid-free paper
10 9 8 7 6 5 4 3 2 1

International Standard Book Number: 978-1-4398-0331-8 (Hardback)

Library of Congress Cataloging-in-Publication Data

Bozzano, Marco.
  Design and safety assessment of critical systems / authors, Marco Bozzano, Adolfo Villafiorita.
       p. cm.
  Includes bibliographical references and index.
  ISBN 978-1-4398-0331-8 (hardcover : alk. paper)
  1. Industrial safety. I. Villafiorita, Adolfo. II. Title.

T55.B68 2011
620.8'6--dc22   2010029858

Visit the Taylor & Francis Web site at
http://www.taylorandfrancis.com

and the Auerbach Web site at
http://www.auerbach-publications.com

To Antonio

To Barbara

Contents

Preface

Safety-critical systems—namely, systems whose failure may cause death or injury to people, harm to the environment, or economical loss—are becoming more complex, both in the type of functionality they provide and in the way they are demanded to interact with the environment. Traditionally, safety analysis techniques and procedures are used to identify risks and hazards, with the goal of eliminating, avoiding, or reducing the probability of failure. However, these techniques are often performed manually and hence are a time-consuming activity, itself vulnerable to human error, because they rely on the ability of the safety engineer to understand and to foresee system behavior. The growing complexity of safety-critical systems requires an adequate increase in the capability of safety engineers to assess system safety, encouraging the adoption of formal techniques.

This book is an introduction to the area of design and verification of safety-critical systems, with a focus on safety assessment using formal methods. After an introduction covering the fundamental concepts in the areas of safety and reliability, the book illustrates the issues related to the design, development, and safety assessment of critical systems. The core of the book covers some of the most well-known notations, techniques, and procedures, and explains in detail how formal methods can be used to realize such procedures. Traditional verification and validation techniques and new trends in formal methods for safety assessment are described. The book ends with a discussion on the role of formal methods in the certification process. The book provides an in-depth and hands-on view of the application of formal techniques that are applicable to a variety of industrial sectors, such as transportation, avionics and aerospace, and nuclear power.

Who should read this book. The book is addressed to both researchers and practitioners in the areas of safety engineering and safety assessment who are interested in the application of formal verification in such areas. It can also be of interest to computer scientists and individuals skilled in formal verification who wish to see how these methodologies can be applied for safety assessment of critical systems. The book can also be used as a reference book for (bachelor and master) students in engineering and computer science.

Prerequisites. The book is mostly self-contained and should be generally accessible to people who have a basic background in mathematics or computer science at the level corresponding to the early years of university courses in engineering or computer science. A prior exposure to topics such as propositional logic, automata theory, model checking, and probability theory could be desirable, although not indispensable.

Structure of the book. The book is structured as follows:

- **Chapter 1, Introduction,** introduces and motivates the main topics of the book.
- **Chapter 2, Dependability, Reliability, and Safety Analysis,** looks in detail at some of the most typical safety criteria that pertain to the design and assessment of safety-critical systems. We start by introducing some common terminology and continue by presenting some fault models and the approaches to dealing with faults, namely fault, detection, fault prediction, fault tolerance, and fault coverage.
- **Chapter 3, Techniques for Safety Assessment,** introduces the traditional notation and techniques for safety assessment. Starting with the definition of hazard and accident, we continue by presenting fault trees, FMECA, HAZOP, Event Tree Analysis, Risk Analysis, and Risk Measures.
- **Chapter 4, Development of Safety-Critical Applications,** looks at the development process of safety-critical systems, by highlighting those management and organizational aspects that most influence the development of safety-critical systems. In this chapter we present a generic development approach that is inspired by various development standards in both the civil and military sectors.
- **Chapter 5, Formal Methods for Safety Assessment,** is an in-depth presentation of formal methods and the role they play in the verification and validation of safety-critical systems. We divert from the "traditional" approach related to the usage of formal methods and propose how formal methods can be effectively used to automate safety analysis techniques.
- **Chapter 6, Formal Methods for Certification,** describes some widely adopted standards for the certification of safety-critical systems. We start with the certification process of aircraft systems and continue by describing how formal methods can be applied to support certification activities.
- Finally, the appendices describe the NuSMV model checker and the FSAP platform, and provide some more references and starting points for further development.

Additional information, including the code for the examples presented in this book, can be retrieved from the web site http://safety-critical.org.

Acknowledgments

The authors would like to thank and acknowledge all the people who provided feedback and support for (earlier versions of) this book, namely, Andreas Lüdtke, Andrea Mattioli, Matteo Negri, Chris Papadopoulos, and Komminist Weldemariam. Special thanks go to Viktor Schuppan for giving specific advice on Chapter 5 and to our editor, John Wyzalek, and all the staff at Taylor & Francis for their help and support.

Finally, Marco Bozzano would like to dedicate the book to his brother Antonio, who prematurely died in November 2004.

Adolfo Villafiorita wishes to thank his family for all the help and support they provided. A big thank-you to his wife Barbara, his dad Enzo, and Andrea, Ombretta, and Rienzo. A special dedication goes to his mom, Geni, who could not see this book completed.

Marco Bozzano

Adolfo Villafiorita

About the Authors

Marco Bozzano is a senior researcher in the Embedded Systems Unit of Fondazione Bruno Kessler, Italy. He has strong expertise in the application of formal methods, and he has published a number of papers in the area of formal verification of safety-critical systems.

Adolfo Villafiorita is a senior researcher at Fondazione Bruno Kessler. He has many years of experience in the application of formal methods in technology transfer projects and in the development of security and safety-critical applications. He is a contract professor at the University of Trento.

Chapter 1

Introduction

1.1 Complex Safety-Critical Systems

Every journey has a start. Ours is the definition of complex safety-critical systems, given in SAE (1996), a set of guidelines for the development of avionic systems:

> "**A complex safety-critical system** is a system whose safety cannot be shown solely by test, whose logic is difficult to comprehend without the aid of analytical tools, and that might directly or indirectly contribute to put human lives at risk, damage the environment, or cause big economical losses."

The definition is peculiar, as it puts together two concepts—namely, *complexity* and *criticality*—that can be defined independently. The motivation for presenting them together in SAE (1996) is obvious: airplanes are both complex and critical. We use this definition for the following reasons:

1. There is a steady trend toward the use of digital systems of increasing complexity in safety-critical applications. Systems need not be digital or complex to be safety critical: The Wright brothers invented and flew the airplane 12 years before Alan Turing was born. However, the flexibility and performance of digital technologies have greatly contributed to increasing their adoption in the safety-critical sector.
2. Systems that are both complex and critical represent an engineering challenge with which traditional techniques have difficulties dealing. Citing Lyu (1996): "The demand for complex hardware/software systems has increased more rapidly than the ability to design, implement, test, and maintain them."

A more detailed discussion and some data will help clarify and put them in perspective.

1.1.1 A Steady Trend toward Complexity

One of the most effective descriptions of the impact that digital technologies have had and are still having on system engineering is given in Brooks (1995), a seminal book about software project management:

> No other technology since civilization began has seen six orders of magnitude price-performance gain in 30 years. In no other technology can one choose to take the gain in **either** improved performance **or** in reduced costs.

Such a trend, first noted by Moore (1965) and since then reported numerous times[1], not only has promoted a capillary diffusion of digital control systems, but has also been a key enabler for the delivery of systems with more functions and increased complexity. Let us provide some more details about both impacts.

The reduction in costs is increasing diffusion. According to Ebert and Jones (2009), in 2008 there were some 30 embedded microprocessors per person in developed countries, with at least 2.5 million function points of embedded software. (A function point is a measure of the size of a software system. Tables to convert function points to lines of source code are available. For instance, Quantitative Software Management, Inc. (2009) estimates one function point as 148 lines of C code.) Millions of these embedded microprocessors are used for safety-critical applications and many of them have faults. For instance, between 1990 and 2000, firmware errors accounted for about 40% of the half-million recalled pacemakers (Maisel et al., 2001; Ebert and Jones, 2009).

The gain in performance is increasing complexity. A recent report by the Jet Propulsion Laboratory (Dvorak, 2009) analyzes the size of NASA flight software for both human and robotic missions. Data for robotic missions are shown in Figure 1.1, where the *x*-axis shows the year and the name of the mission and the *y*-axis shows the flight software size, using a logarithmic scale.

As can be seen from the diagram, software size is growing exponentially. The same trend is shown by NASA manned missions (Apollo, Shuttle, and International Space Station), although the number of data points is too small to demonstrate any trend, as pointed out by the authors of the report. Similar growth can also be observed in other domains, such as civil and military avionics, automotive, and switching systems, to name a few (Ferguson, 2001; Dvorak, 2009; Ebert and Jones, 2009).

---

[1] As an example, we quote from Air Force Inspection and Safety Center (1985), written 25 years ago: "The development and usage of microprocessors and computers has grown at a phenomenal rate in recent years, from 1955, when only 10 per cent of our weapon systems required computer software, to today, when the figure is over 80 per cent."

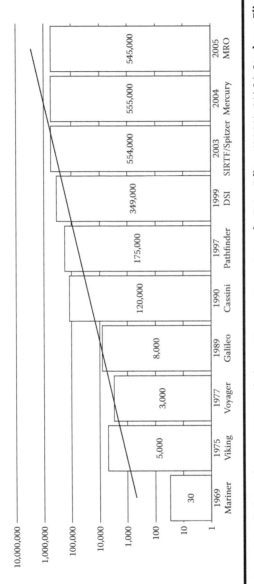

Figure 1.1 Growth in flight software, NASA missions. (Source: From Dvorak, D.L., Editor (2009). NASA Study on Flight Software Complexity. Available at http://oceexternal.nasa.gov/OCE_LIE/pdf/1021608main_FSWC_Final_Report.pdf.)

Together with size, complexity is also increasing. This is due to various factors, among which we mention:

- **Number of functions.** In the words of Dvorak, (2009), software is working as a sponge "because it readily accommodates evolving understanding, making it an enabler of progress." A very good example of this phenomenon are jet-fighters. Modern jet-fighters, in fact, are designed to be slightly aerodynamically unstable. That is, a small variation in the current flight conditions causes the plane to abruptly change trajectory. This feature allows *fast transients*, that is, quick changes in speed, altitude, and direction—maneuvers on which the life of a pilot in combat might depend. However, to maintain the flight level, the airplane must be kept under active and constant control. This is performed by a digital system (called fly-by-wire, FBW), because the precision and the frequency of corrective actions that must be taken make the task impossible for a pilot. The FBW, thus, continuously reads plane's data and pilot's commands and constantly positions the actuators according to the data received, in practice decoupling the pilot from the controls of the plane. See, for example, Langer et al. (1992), NASA (2009), and Various Authors (2009) for more details.
- **Number of states.** Software and, more generally, digital systems, have a large number of states that can make their comprehension difficult and exhaustive testing impossible.
- **Discrete behavior.** Software models discrete systems with discontinuous behaviors. That is, a little variation in one program input could cause a great variation in one output. Such discontinuities are, in principle, uniformly distributed over the whole input space. The consequence is that performing a test on an input value tells little about the behavior of the system on the neighboring region. This is rather different from analog systems, in which small variations in one input usually cause small variations in outputs.
- **Invisibility.** As pointed out in Brooks (1995), software is invisible, unvisualizable, and its reality is not embedded in space; software can be visualized only by overlapping several different views (e.g., data-flow, control-flow) that define its behavior. Brook's observation holds up today. The UML, a standard notation in object-oriented programming, defines nine different diagrams to model a software system. At least five of them must be used to properly model a system's architecture (Kruchten, 1995).

1.1.2 An Engineering Challenge

Safety-critical systems have stringent requirements. Not all systems are equally critical when they fail. Looking at the consequences associated with system or function failure is thus a good way to discriminate among the different levels of criticality.

Such a classification can be found, for instance, in Sommerville (2007), where the author distinguishes among:

■ Business-critical systems
■ Mission-critical systems
■ Safety-critical systems

In the first case (business-critical systems), a failure of the system might cause a high economic loss. This is often due to the interruption of service caused by the system being unusable. Examples of business-critical systems are a stock-trading system, the ERP system[2] of a company, or an Internet search engine, such as Google[3].

In the second case (mission-critical systems), failures might cause the loss of a function necessary to achieve one of the goals for which the system was designed. Examples of mission-critical systems are an unmanned spacecraft and the software controlling a baggage-handling system of an airport.

In the third case (safety-critical systems), system failures might cause risk to human life or damages to the environment. Examples of safety-critical systems are aircraft, the controller of an unmanned train metro system, and the controller of a nuclear plant.

Critical systems are further distinguished between fail-operational and fail-safe systems, according to the tolerance they must exhibit to failures. In particular:

■ **Fail-operational** systems are typically required to operate not only in nominal conditions—that is when all the (sub)components of the system work as expected—but also in degraded situations, that is, when some parts of the system are not working properly. Airplanes are fail-operational because they must be able to fly even if some components fail.
■ **Fail-safe** systems are demanded to safely shut down in case of single or multiple failures. Trains are fail-safe systems because stopping a train is typically sufficient to put it into a safe state.

(Safety-critical) systems fail for the most diverse reasons. Following O'Connor (2003), we mention the following causes for systems to fail:

■ **The design might be inherently incapable.** This is typically due to errors during development that result in building a system that it is inadequate for the purpose for which it was devised. Causes are the most diverse. Some essential requirement might have been missed during the specification, such as some environmental condition the system should have been resilient to. Some error

---

[2] Enterprise Resource Planning system.
[3] Notice that such an interruption would be both a loss for the search engine provider (e.g., missed revenues in advertisement) and its users (e.g., decreased productivity in finding information on the Internet).

might have been introduced during design, as it is so common in software, for instance. We also mention sneaks, integration, and interaction errors that occur when the system does not work properly even though all its components do.

■ **The system might be overstressed.** This is often due to the system being operated in environmental conditions for which it was not designed. Notice that in complex systems, stress on a component might be caused by other components failing in unexpected ways. For instance, a short circuit in an electronic component might cause an excessive voltage in another one.

■ **Variation in the production and design.** This is due to natural variations in the materials, in the production processes, and in quality assurance procedures. Problems arise when a component that is, let us say, less resistant than average is subject to a load that is above the average.

■ **Wear-out and other time-related phenomena.** All components become weaker with use and age. As a consequence, their probability of failure increases over time. The problem can be mitigated by proper system maintenance and replacement of components before they wear out. However, environmental conditions (e.g., mechanical vibrations in rockets, pressure in submarines), design issues (e.g., friction on the insulation of an electric cable installed in the wrong position) can accelerate wear-out in unpredicted ways. Moreover, for certain systems, such as spacecraft, replacement may be impossible.

■ **Errors.** Errors can occur in any phase of a system's life cycle. We mentioned above errors occurring during system specification and development. Errors can occur also during maintenance (e.g., a component replaced in the wrong way), during operations (e.g., due to problems in training, documentation, or just distraction), or during system disposal.

An Engineering Challenge. The development of critical systems thus adds a further level of complexity to standard engineering activities because it requires to consider, and properly deal with, all the diverse causes of failure, so that the system can maintain a function even if some components fail or operators make errors.

This requires the adoption of development processes in which safety is considered from the early stages. In aeronautics, for instance, safety requirements (i.e., requirements stating the (degraded) conditions under which systems must remain operational) are defined along with the other system requirements. During system engineering, development activities are conducted in parallel with a set of safety analysis activities that have the specific goal of identifying all possible hazards, together with their relevant causes, in order to assess if the system behaves as required under all the operational conditions. These activities are crucial (e.g., for system certification) to ensure that the development process is able to guarantee the specific safety level assigned to the system.

1.2 Dealing with Failures: A Short History of Safety Engineering

System safety has been a concern in engineering for a long time and it has always been an important factor in determining systems' adoption. Elevators were in use since the third century, but they became popular only after 1853 when Elisha Otis demonstrated a freight elevator equipped with a safety device to prevent falling, in case a supporting cable should break (History of the Elevator, 2009).

Safety engineering, however, matured as a discipline only in the past 50 years. In the following few paragraphs we provide an overview of the main steps that led to this revolution. See Air Force Safety Agency (2000), Ericson (1999), Leveson (2003), and Ericson (2006) for more details.

System safety, as we know it today, is strictly related to the problems the U.S. Air Force experienced with accidents after World War II and its efforts to prevent them. According to Hammer (2003), from 1952 to 1966 the U.S. Air Force lost 7,715 aircraft. In the accidents, 8,547 persons were killed. Although many of those accidents were blamed on pilots, there were many who did not believe the cause was so simple (Leveson, 2003).

During the 1950s, a series of accidents involving the Atlas ICBM contributed to a growing dissatisfaction with the "fly-fix-fly" approach. At the time, safety was not a specific system engineering activity, but rather a concern distributed among the project team. After system deployment, if an accident occurred, investigations reconstructed the causes to allow engineers to "fix" the design and prevent future similar events.

The approach, however, soon became ineffective because it did not help prevent accidents with causes different from those investigated, and deemed too costly and too dangerous, considering, for example, the risks of an accident involving a nuclear weapon. These considerations eventually led to abandoning the existing development practices and adopting, instead, an approach in which system safety activities are integrated into the development process (Leveson, 2003). Such an integrated approach had its roots in a seminal paper by Amos L. Wood, "The Organization of an Aircraft Manufacturer, Air Safety Program," presented in 1946, and in a paper by William I. Stieglitz, "Engineering for Safety," published in 1948 (Air Force Safety Agency, 2000). From Stieglitz, H.A. Watson of Bell Laboratories first conceived Fault Tree Analysis, in connection with the development of the Launch Control System of the Minuteman missile. The technique proved so successful that it was later extensively applied to the entire Minuteman program.

In 1965, Boeing and the University of Washington sponsored the first System Safety Conference and later developed a software system for the evaluation of multiphase fault trees. The technique soon caught on in other areas, most notably the civil nuclear sector, which has been, since then, a great contributor to the technique and to safety in general. After the Apollo 1 launchpad fire in 1967, NASA hired

Boeing to implement an entirely new and comprehensive safety program for the Apollo project. As part of this safety effort, Fault Tree Analysis was performed on the entire Apollo system (Ericson, 1999). The technique was finally consolidated with the release by NUREG of the *Fault Tree Handbook* (Vesely et al., 1981).

Software safety analysis also had its roots in the 1960s. The first references are dated 1969 and, since then, the subject has gained momentum and interest. We cite the Air Force Inspection and Safety Center (1985):

> "Software safety, which is a subset of system safety, is a relatively new field and is going to require a conscientious effort by all those involved in any system acquisition process or development effort to insure it is adequately addressed during system development."

Finally, in recent years, military standards such as MIL-STD-1574A (eventually replaced by the MIL-STD-882 series) and the growing demand for safety in civil applications—especially in the nuclear and transportation sector—have greatly contributed to the standardization of techniques, on the one hand, and to the standardization of development processes of safety-critical applications, on the other.

1.3 The Role of Formal Methods

As highlighted by Bowen and Stavidrou, lives have depended on mathematical calculations for centuries. In the nineteenth century, errors in logarithmic tables caused ships to miscalculate their position and possibly wreck as a result of such errors (Bowen and Stavridou, 1992). Mathematical representations of hardware and software systems (formal methods) have emerged in the past 30 years as a promising approach to allow a more thorough verification of the system's correctness with respect to the requirements, using automated and hopefully exhaustive verification procedures.

As described earlier, safety-critical systems are becoming more complex, both in the type of functionality they provide and in the way they are required to interact with their environment. Such growing complexity requires an adequate increase in the capability of safety engineers to assess system safety, a capability that is only partially matched by the progress made in the use of traditional methodologies, such as Fault Tree Analysis and failure mode and effect analysis, often carried out manually on informal representations of systems. The use of formal techniques for the safety assessment of critical systems, however, is still at a relatively early stage. This is due to the following reasons:

- **The role of formal methods for system design.** Nearly all standards used as references for the development and certification of safety-critical systems make little mention, if at all, of formal methods. Main causes include the maturity of techniques and tools when the standards were issued, skills needed for properly mastering the techniques, and difficulties related to an effective integration of formal methods in the development process.

■ **The role of formal methods for system safety assessment.** Formal methods have traditionally been used to support system verification activities. There is, however, a fundamental difference between system verification and safety activities. The first is meant to demonstrate that the *nominal* system works as expected. A *single* counterexample is sufficient to show that a requirement is violated. The second is meant to support design by demonstrating that the *degraded* system works as expected. To do so, it is necessary to highlight *all possible combinations of failures* that lead to losing a function. This requires significant changes, both in the way in which systems are used and in the way verification engines are implemented.

■ **Integration between design and safety assessment.** The information linking the design and the safety assessment phases is often carried out informally, and the connection between design and safety analysis may be seen as an *over-the-wall process*. Quoting Fenelon et al. (1994), "A design is produced with some cognisance of safety issues, it is 'tossed over the wall' to safety assessors who analyse the design and later 'toss it back' together with comments on the safety of the design and requests for change. Whilst something of a caricature, the shove is not entirely unrepresentative of current industrial processes." Thus, even when formal methods are used to support design activities, the extra effort spent there cannot be reused for safety assessment, because the formal designs are "lost" by this lack of integration between activities.

Recent developments are making the use of formal methods for system verification more appealing. We mention improvements on the representational power of formal notations, increased efficiency of the verification engines, better tool support, and significant improvements in the ease of use. Steady progress has also been measured with respect to the usability of formal methods for safety assessment. Novel algorithms, based on model checkers, have been defined, for instance, to support the automatic computation of fault trees and to automate common cause analysis; see, for example, Bozzano et al. (2003), Joshi et al. (2005), Åkerlund et al. (2006), Bozzano and Villafiorita (2007), and Bozzano et al. (2007).

Despite the progress mentioned above, we are still far from a full, top-down, and completely automated verification of (complex) safety-critical systems. Formal methodologies, however, represent one of the most promising areas to improve the quality and safety of complex systems.

1.4 A Case Study: Three Mile Island

The complexity of the environment, functions performed, and difficult-to-understand interactions among system parts when components fail are main causes of engineering failures and accidents. In this section we briefly describe the Three

Mile Island accident, one of the worst in the civil nuclear sector. We do so by also presenting a formal model of the plant, written in the input language of NuSMV, a symbolic model checker. The formal model will be used in this book to reproduce the accident and demonstrate some of the techniques used for safety assessment. (See Appendix A for a description of the NuSMV model checker.)

The formal model has been built using the know-how obtained from an analysis of the accident. We cannot therefore pretend, nor do we claim, that the model and formal methods could have been used to build a better plant and prevent the accident.

Nevertheless, the example is a very good exercise for the following reasons:

1. It shows how we can use a formal language, whose expressiveness is limited to finite state machines over discrete data types, to model a physical system in which the behavior is determined by the laws of thermodynamics and nuclear physics. The challenge is to provide a suitable level of abstraction that allows us to accurately model the qualitative behaviors necessary for the analyses we want to perform. This is very common when using formal methods and model checking, as abstraction is often the only way to keep analyses manageable.

2. It shows how we can model a complex physical system using functional blocks. The trick here is to use the blocks to encode the "flow of information" (e.g., a pump is pumping) rather than the "flow of physical elements" (e.g., the coolant flows from the core to the steam generator). This results in a significant difference between the physical structure of the plant (at the level of abstraction considered in this book) and the actual functional block model. The approach is quite common when using model checkers for physical systems.

3. It presents an example in which safety analyses (and, in particular, fault trees) are conducted by observing the dynamics of the system rather than by statically analyzing the system architecture. This is a noteworthy improvement over standard assessment techniques and shows one of the advantages that can be obtained with the adoption of formal methods for safety assessment.

Most of the work presented here can be readily applied to other modeling languages and verification systems. For instance, to see the same example specified using the formalism of safecharts, have a look at Chen (2006).

1.4.1 Pressurized Water Reactors (PWRs)

Pressurized water reactors (PWRs) are second-generation nuclear power plants conceived in the 1950s and used for a variety of applications, ranging from the propulsion of nuclear submarines to the production of electricity for civil use. PWRs are the most common type of nuclear power plant, with hundreds of systems used for naval propulsion and more than 200 civil installations in Japan, the United States, France, and Russia (World Nuclear Association, 2009; Various Authors, 2007).

The electricity in a PWR is generated using two independent hydraulic circuits that transfer the heat produced by the nuclear reaction to a turbine. The coolant used in the circuits is water. The adoption of two independent circuits is a design feature that PWRs share with other kinds of reactors, such as the pressurised heavy water reactor (PHWR) and the advanced gas-cooled reactor (AGR), that, however, use different coolants.

In the first circuit, called the **primary circuit**, the water is directly in contact with the nuclear material or **fuel**, from now on. The coolant in the first circuit is highly pressurized, so that it can remain in its liquid form at the high temperature produced by the fuel (about 300°C).

The primary circuit has two uses. First, it keeps the reaction under control. Second, it transfers the heat generated by the fuel to the coolant of the second circuit, through a **steam generator**. In the steam generator, in fact, the high-temperature water of the primary circuit is cooled by the low-temperature water of the second circuit. This, in turn, vaporizes to steam. The steam thus generated in the secondary circuit activates a turbine that produces electricity. A third circuit is used to cool the liquid of the secondary circuit after it has flowed through the turbine. Pumps ensure that the fluids keep flowing in all circuits. See Figure 1.2 for a high-level sketch of the plant.

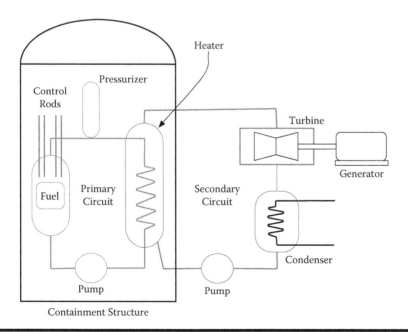

Figure 1.2 High-level sketch of a pressurized water reactor.

The nuclear reaction is controlled by three means:

1. The *coolant of the first circuit*. In fact, to keep the nuclear reaction going, the neutrons emitted by the fuel fission need to be slowed so that they can generate more fission. This is accomplished by a **moderator**. Moderators typically used include water, heavy water, and graphite. In a PWR, the moderator is the water of the first circuit itself. Among the advantages of such a choice, the fact that the capacity of water to slow neutrons decreases as its temperature increases. Thus, the water in the PWR has a negative feedback on the reaction, as any increase in the nuclear reaction causes the temperature of the water to increase, which eventually slows the reaction[4].
2. **Chemicals,** such as borate, that absorb neutrons. The chemicals are injected in the first circuit and reach the fuel, thus slowing the reaction. Chemicals can be used as a secondary shutdown system.
3. **Control rods.** These are rods of a neutron-absorbing material, such as cadmium. The rods can be inserted or withdrawn from the fuel to control the rate of the reaction. Full insertion of the rods in the core causes the nuclear reaction to stop. In an emergency situation, the control rods can be completely inserted in the core in a matter of seconds, in an operation called SCRAM[5].

To complete our 10,000-foot overview of a PWR, we need to mention two other components, the **coolant tanks** and the **pressurizer**, that ensure that the pressure in the first circuit remains within safe bounds. In fact, if the pressure is too low, bubbles (also called steam cavities) start forming in the circuit, decreasing the cooling effect and wearing components, such as the pumps. If the pressure is too high, the circuit goes **solid**, that is, the coolant does not have the tiniest space for any expansion. In such a situation, any further increase in the temperature of the coolant might cause the explosion of the circuit.

For these reasons, the pressurizer is equipped with various safety devices. A special valve called the pilot-operated relief valve (PORV) can be used to release some of the coolant of the circuit. Vice versa, the coolant tanks can be used to increase the pressure in the circuits by pumping more coolant or enriching the water of the first circuit with borate.

---

[4] This is the opposite behavior of other types of nuclear power plants, such as the light-water, graphite-moderated reactor in use at Chernobyl. In these plants, an increase in the temperature of the moderator does not have any moderation effect. On the contrary, temperature is a source of instability, since any increase in nuclear activity actually reduces the effect of the moderator, which, in turn, could cause more fission to occur.

[5] According to NUREG, SCRAM is an acronym for "safety control rod axe man," the worker responsible for inserting the emergency rods on the first reactor in the United States.

1.4.2 When Things Go Wrong

Controlling the circuits is essential to safely operate a PWR. For one thing, the coolant must be kept flowing in all the circuits, so that heat is properly taken away from the reactor. Even more importantly, the controller must maintain the pressure of the primary circuit within bounds, by acting on the PORV or on the coolant tanks, if any deviation from the standard values occurs. The main risk is uncovering the core, either because the circuit goes "solid" and explodes, or because the core is not covered by water anymore and melts. These are the most feared accidents, as "in a worst-case accident, the melting of nuclear fuel would lead to a breach of the walls of the containment building and release massive quantities of radiation to the environment" (U.S. Nuclear Regulatory Commission, 2009).

On March 28, 1979, at Three Mile Island, Unit 2, maintaining the circuits under control turned out to be very difficult, rather stressing, and not without consequences. In fact, due to a sequence of failures and a confusing user interface, fluid was drained from the primary circuit until the core was partially uncovered and partially melted. Even though the accident could have caused damage similar to that of Chernobyl, the situation was eventually contained and no serious damage to people or to the environment occurred.

The sequence of events is roughly as follows. Small plastic debris introduced by mistake during maintenance into the secondary circuit caused its pumps to stop. As a consequence, heat started accumulating in the primary circuit. To release the excess pressure that was building up in the primary circuit, the PORV was opened. As expected, the pressure dropped to a safe value. The operator commanded the PORV valve closed. Here is where things started going wrong.

The PORV failed to close and the coolant kept spilling from the circuit. Problems in the user interface (e.g., it showed the command issued to the valve rather than its actual status), the physics of the system (the level of the coolant in the pressurizer, used as an indicator of the pressure of the primary circuit, kept raising, as a result of the leak), and other contributing factors led the operators to believe that the pressure in the circuit was still rising.

The engineers thus stopped the emergency pumps, which had automatically started to inject more fluid in the primary circuit, and ordered open the relief valves of the primary circuit, to avoid what they believed was happening in the circuit, namely, going solid[6]. In fact, the opposite was occurring. When the fluid in the primary circuit became too low, the core started melting. Eventually the problem was sorted out, the PORV valve cut off, and the leak stopped. The water level in the primary circuit was restored.

---

[6] The fact that the emergency pumps had autonomously started without reason on earlier occasions contributed to misleading the operators.

Subsequent events, such as a small release of radioactivity in the environment and the build-up of a large hydrogen bubble in the containment chamber raised further concerns in the 2 days after the accident. The unit was eventually shut down and has not operated since then. According to the Senate report on the accident, no damage to people or the environment resulted (TMI, 1979; Rogovin and Frampton, 1980). Unit 1 of Three Mile Island, of similar design, and already in operation by the time of the accident at Unit 2, has kept working since its deployment in the 1970s.

This description of the accident does not give an accurate account of the confusion and tension at the time of the accident. Suffice it to say that between 20 and 60 operators were present in the control room at the time of the accident, and that the level of heat and radiation in the core became so high that, following Nuclear Regulatory Commission regulations, a general emergency had to be declared. Major damage was caused to the reactor: some 20 tonnes of molten radioactive material spilled into the bottom of the reactor vessel. Had the situation not been brought back under control, the radioactive material would have eventually spilled out.

Several reports, books, and web sites provide more detailed and accurate information on the matter; see, for instance, Knief (1992), Walker (2004), and Chiles (2002). Finally, the U.S. Nuclear Regulatory Commission (2009) and the Smithsonian National Museum of American History, Behring Center (2009) are excellent starting points and pointers to other material on the Internet.

1.4.3 The Plant Structure

Figure 1.3 shows the schematic of a PWR that we use for our formal model. The picture is not meant to depict the actual layout, whose complexity is far higher than shown, but rather to provide a relatively complex and sufficiently detailed example. On the left hand side we see the *reactor* and the primary (first) circuit.

The fluid in the circuit keeps moving by means of two pumps (*P1a* and *P1b*). A *pressurizer* and the *PORV* (top left side of the figure) ensure that the pressure does not become too high. The *PORV* can be excluded by a valve *BV1*. A *safety valve* automatically intervenes if the pressure is too high. Coolant can also be extracted from the primary circuit using a second valve *BV2*. Two coolant tanks (bottom left part of the figure) can be used to inject coolant into the primary circuit and increase its pressure. To do so, the valve must be open and the pump started.

The secondary circuit is equipped with two pumps (*P2a* and *P2b*) that keep the coolant flowing into the steam generator. The steam generated in the steam generator *C2* moves the rotor of the turbine, which generates electricity. Finally a condenser *C3* uses the coolant of the third circuit to condense the steam of the second circuit so that it can be pumped back to *C2*.

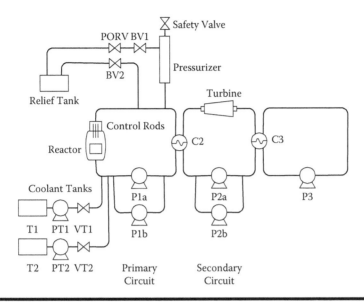

Figure 1.3   A schematic view of a PWR.

1.4.4 Formal Model of a PWR

NuSMV specifications are based on functional blocks, or *modules* using NuSMV terminology, the behavior of which is specified by state machines and that are combined to build complex specifications.

Our model of the PWR is based on the following modules:

- **Pump.** Pumps might either be working (pumping the coolant) or not. The inputs to the pump are the command issued by the operator (i.e., either `start` or `stop`), and the status of the coolant (`normal`, `steam_cavities`, `solid`). Steam cavities in the coolant, might, in fact, cause the pump to break, as changes in the density of the coolant cause sudden accelerations and decelerations that overstress the pump.
- **Valve.** The valve is modeled as a component that has as input the command received by the operator (i.e., either `open` or `close`), and produces as output its state, (i.e., either `opened` or `closed`).
- **Circuit.** Circuits transfer heat from one source to another. If we assume that the conductivity of the coolant is perfect (so that, e.g., there are no losses between input and output), then we can model this behavior with a functional block composed of two "wires," that transfer their inputs to their outputs. This is depicted in Figure 1.4 (upper part), where, on the left-hand side, we show how heat "flows" in a circuit and, on the right-hand side, we present the corresponding functional block.

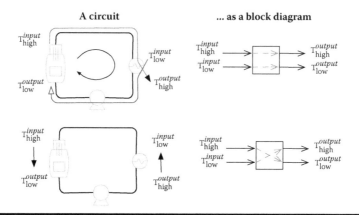

Figure 1.4   The representation of a circuit as a functional block.

This is not enough, however, as the heat transfer is guaranteed only if the pump is working. When the coolant does not flow in the circuit, the temperature of the coolant eventually matches that of the source with which it is in contact. (We, in fact, are assuming that both the "low" and the "high" sources have infinite capacity.) In such a case, therefore, the effect on the functional representation is that of a "short-circuit" (see Figure 1.4), in which the output produced depends on the nearest input source.

We need to model two other features that are necessary for the primary circuit, namely, the level and the status of the coolant. For the first, we assume fixed input and output capacities, so that the level of coolant depends on the status of the input and output valves. For the latter, we define three states, normal, steam_cavities, and solid, that represent, respectively, a normal situation, the fact that steam cavities have formed, and the fact that the coolant in the circuit is "solid." The status depends on the level of the coolant and the temperature measured at either side of the circuit.

■ **Reactor.** In our simplified model, the reactor is a source of heat that is controlled by two factors, namely, the position of the control rods and the level of the fluid in the reactor. The control rods have three positions, extracted, partially_inserted, and inserted, resulting from corresponding operator commands, extract, partially_insert, and insert. The coolant has the effect of moderating the reaction. We do not model mitigation effects due to the injection of chemicals.

The other components of the plant illustrated in Figure 1.3 do not need to be modeled. We do not model the turbine, because the amount of energy produced is proportional to the energy of the coolant in the second circuit and is completely irrelevant for the analyses we want to perform. We do not model the condenser

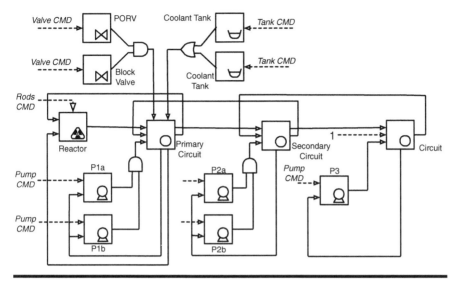

Figure 1.5   The functional block diagram of the example.

and the steam generator, because their behavior is "embedded" in the model of the circuit. Finally we do not model tanks because we assume them to be of infinite capacity. Notice that the models of all these components might be essential for other kinds of analyses or in different settings.

Figure 1.5 shows the resulting block diagram. The functional block diagram is the basis for the formal NuSMV model, because each block is translated into an NuSMV MODULE, and the behavior of each module is specified through a state machine. We do not show the source code here: it can be found at http://safety-critical.org.

The functional block diagram and the executable specification will be used in the next chapters to present various manual and automated safety assessment techniques.

References

Air Force Inspection and Safety Center (1985). Software System Safety. Technical Report AFISC SSH 1-1. Kirkland Air Force Base, NM: Air Force Inspection and Safety Center.

Air Force Safety Agency (2000).   Air Force System Safety Handbook.   Available at http://www.system-safety.org/Documents/AF_System-Safety-HNDBK.pdf. Last retrieved on November, 15, 2009.

Åkerlund, O., P. Bieber, E. Böede, et al. (2006). ISAAC, a framework for integrated safety analysis of functional, geometrical and human aspects. In *Proc. European Congress on Embedded Real Time Software (ERTS 2006)*.

Bowen, J.P. and V. Stavridou (1992). Safety-critical systems, formal methods and standards. *BCS/IEE Software Engineering Journal* 8(4), 189–209.

Bozzano, M., A. Cimatti, and F. Tapparo (2007). Symbolic fault tree analysis for reactive systems. In *Proc. 5th International Symposium on Automated Technology for Verification and Analysis (ATVA 2007)*, Volume 4762 of *LNCS*, pp. 162–176. Berlin: Springer.

Bozzano, M. and A. Villafiorita (2007). The FSAP/NuSMV-SA safety analysis platform. *Software Tools for Technology Transfer* 9(1), 5–24.

Bozzano, M., A. Villafiorita, and O. Åkerlund et al. (2003). ESACS: An integrated methodology for design and safety analysis of complex systems. In *Proc. European Safety and Reliability Conference (ESREL 2003)*, pp. 237–245. Leiden, The Netherlands: Balkema.

Brooks, F. P. (1995). No silver bullet—essence and accident. In *The Mythical Man Month (Anniversary Edition with four new chapters)*, Chapter 16. Reading MA: Addison-Wesley.

Chen, Y.-R. (2006). Automatic Failure Analysis Using Extended Safecharts. Master's thesis, Institute of Computer Science and Information Engineering College of Engineering, National Chung Cheng University.

Chiles, J.R. (2002). *Inviting Disaster—Lessons from the Edge of Technology*. New York: Harper Business.

Dvorak, D.L., Editor (2009). NASA Study on Flight Software Complexity. Available at http://www.nasa.gov/offices/oce/documents/FSWC.study.html.

Ebert, C. and C. Jones (2009). Embedded software: Facts, figures, and future. *Computer* 42(04), 42–52.

Ericson II, C.A. (1999). Fault tree analysis—A history. In *Proc. 17th International System Safety Conference*.

Ericson II, C.A. (2006). A short history of system safety. *Journal of System Safety* (eEdition) 42(3).

Fenelon, P., J.A. McDermid, M. Nicolson, and D. Pumfrey (1994). Towards integrated safety analysis and design. *SIGAPP Applied Computing Review* 2(1), 21–32.

Ferguson, J. (2001). Crouching dragon, hidden software: Software in DoD weapon systems. *IEEE Software* 18(4), 105–107.

Hammer, W. (2003). *Product Safety Management and Engineering (2nd ed.)*. Des Plaines, IL: American Society of Safety Engineers.

History of the Elevator (Last retrieved on November 15, 2009). The History of the Elevator. Available at http://inventors.about.com/library/inventors/blelevator.htm.

Joshi, A., S. Miller, M. Whalen, and M. Heimdahl (2005). A proposal for model-based safety analysis. In *Proc. 24th Digital Avionics Systems Conference (DASC 2005)*. Washington, D.C.: IEEE Computer Society.

Knief, R.A. (1992). *Nuclear Engineering: Theory and Technology of Commercial Nuclear Power*. London: Taylor & Francis.

Kruchten, P. (1995). Architectural blueprints—the "4+1" view model of software architecture. *IEEE Software* 12(6), 44–50.

Langer, D., J. Rauch, and M. Rössler (1992). Fly-by-wire systems for military high performance aircraft. In M. Schiebe and S. Pferrer (Eds.), *Real-Time Systems Engineering and Applications*, pp. 369–395. Dordrecht, The Netherlands. Kluwer Academic.

Leveson, N.G. (2003). White Paper on Approaches to Safety Engineering. Available at http://sunnyday.mit.edu/caib/concepts.pdf. Last retrieved on November 15, 2009.

Lyu, M.R. (Ed.) (1996). *Handbook of Software Reliability Engineering*. Washington, D.C.: IEEE Computer Society, and New York: McGraw-Hill.

Maisel, W., M. Sweeney, W. Stevenson, K. Ellison, and L. Epstein (2001). Recalls and safety alerts involving pacemakers and implantable cardioverter-defibrillator generators. *Journal of the American Medical Association* 286(7), 793–799.

Moore, G.E. (1965). Cramming more components onto integrated circuits. *Electronics* 38(8), 114–117.

NASA (Last retrieved on November, 15, 2009). F-8 digital fly-by-wire aircraft. Available at `http://www.nasa.gov/centers/dryden/news/FactSheets/FS-024-DFRC.html`.

O'Connor, P.D. (2003). *Practical Reliability Engineering* (4th ed.). New York: Wiley.

Quantitative Software Management, Inc. (Last retrieved on November 15, 2009). Function Point Languages table. Available at `http://www.qsm.com/?q=resources/function-point-languages-table/index.html`.

Rogovin, M. and G.T. Frampton (1980). *Three Mile Island. A Report to the Commissioners and to the Public*, Volume I-II. NUREG/CR-1250.

SAE (1996). Certification Considerations for Highly-Integrated or Complex Aircraft Systems. Technical Report ARP4754, Warrendale, PA: Society of Automotive Engineers.

Smithsonian National Museum of American History, Behring Center (Last retrieved on November 15, 2009). Three Mile Island: The Inside Story. Available at `http://americanhistory.si.edu/TMI/`.

Sommerville, I. (2007). *Software Engineering (8th ed.)*. Reading, PA: Addison-Wesley.

TMI (1979). Report of the President's Commission on the Accident at Three Mile Island. Available at http://www.pddoc.com/tmi2/kemeny/index.html (Last retrieved on November 15, 2009).

U.S. Nuclear Regulatory Commission (Last retrieved on November 15, 2009). Backgrounder on the Three Mile Island Accident. Available at `http://www.nrc.gov/reading-rm/doc-collections/fact-sheets/3mile-isle.html`.

Various Authors (2007). *World Nuclear Industry Handbook*. Kenf, United Kingdom: Nuclear Engineering International. Surrey, England: Business Press International.

Various Authors (Last retrieved on November, 15, 2009). Aircraft Flight Control System: Available at `http://en.wikipedia.org/wiki/Aircraft_flight_control_systems`.

Vesely, W.E., F.F. Goldberg, N.H. Roberts, and D.F. Haasl (1981). Fault Tree Handbook. Technical Report NUREG-0492, Systems and Reliability Research Office of Nuclear Regulatory Research U.S. Nuclear Regulatory Commission.

Walker, J. Samuel (2004). *Three Mile Island: A Nuclear Crisis in Historical Perspective*. Berkeley, CA: University of California Press.

World Nuclear Association (Last retrieved on November 15, 2009). Nuclear Power Reactors. Available at `http://www.world-nuclear.org/info/inf32.html`.

Chapter 2

Dependability, Reliability, and Safety Assessment

2.1 Introduction

In this chapter we look in detail at some of the most typical safety criteria that pertain to the design and assessment of safety-critical systems. First of all, we need to introduce some terminology, in particular the notions of *fault*, *error*, and *failure*. As strange as it may sound, there seems to be no common agreement in the literature on this topic, in that different standards (sometimes even standards issued by the same authority) may adopt different definitions of these basic notions.

Here, following Storey (1996), we define a **fault** as the presence of a defect or an anomaly in an item or system. Notice that a fault may not manifest itself, except under certain circumstances. For instance, a hardware fault causing a valve to remain stuck in the "close" position may not manifest until the valve is eventually commanded to open. In such a case, we say that the fault is **dormant**. An **error** is the way in which a fault manifests itself, that is, it is a deviation of the behavior of an item or system from the required operation. In the previous example, the inability of the valve to open on command is an error that is a consequence of the corresponding fault. Notice that an error is an *event*, as opposed to a fault, which is a property of the state. Finally, a **failure** is defined as the inability of an item or system to perform its required function. When an error has the potential to cause a failure, but it has not yet done so, we say that the error is **latent**. Notice that an error does not always result in a failure; for instance, a tank may be designed with a redundant architecture that uses primary and secondary flow-out valves. In case a fault prevents the primary valve from opening, the secondary valve could be operated to perform the required system function.

Faults can be categorized in several ways. For instance, we can distinguish hardware faults, such as the valve fault of the previous example, and software faults such as a buggy piece of code. Faults can also be categorized depending on their duration; namely, a fault can be permanent or transient. For example, it is possible for a sensor to produce a wrong reading due to transient conditions of the environment in which it is operating (e.g., due to electromagnetic radiation). We further discuss the classification of faults in Section 2.3, whereas in Section 2.4 we discuss how faults can be modeled and provide different examples of *fault models*.

2.2 Concepts

In this section we review some basic concepts related to the development and operation of safety-critical systems. A **safety-critical system** (also called a **safety-related system** or **safety instrumented system**) is a system designed to ensure safe operation of the equipment or the plant that it controls.

Safety-critical systems must typically comply with a number of requirements of different forms. In general, we distinguish between **external requirements** and **internal requirements**. External requirements involve the operation of a given system as perceived by the users; examples of external requirements are correctness, usability, safety, and reliability. Internal requirements involve properties of the system related to its design and maintainability; examples of internal requirements are testability, portability, and reusability. A further classification distinguishes **functional requirements** from **nonfunctional requirements**. For instance, safety and correctness are examples of functional requirements, whereas reliability and availability are examples of nonfunctional requirements. Note that, in general, there may be potential conflicts between different requirements; for instance, system safety could be achieved at the price of reduced availability or reliability. Some of the main notions that are specific to safety-critical systems and their interrelationships are discussed below.

2.2.1 Safety

Safety can be described as a characteristic of the system of not endangering, or causing harm to, human lives or the environment in which the equipment or plant operates. That is, safety evaluates system operation in terms of freedom from occurrence of catastrophic failures.

It is possible to distinguish different forms of safety. In particular, **primary safety** relates to the direct effects that operating the system may cause; for instance, damages due to electric shocks or fire resulting from the operation of a computer's hardware. **Functional safety** relates to the safe operation of the equipment under control, for instance, a computer may cause damage due to a software bug that causes incorrect operation of the system under control. Finally, **indirect safety** relates to the indirect consequences of a failure, for instance, the unavailability of the service that is controlled by a computer system.

Notice that safety is different from **security**. The latter concept primarily refers to privacy or confidentiality issues, and typically considers events such as unauthorized access and malicious events, whereas safety also considers actions that were intended during the system design.

2.2.2 Reliability

Reliability refers to the characteristic of a given system of being able to operate correctly over a given period of time. That is, reliability evaluates the probability that the system will function correctly when operating for a time interval t. Hence, reliability is a function of time (the longer the time interval, the lower the corresponding reliability). Equivalently, reliability can be defined in terms of *failure rate*, that is, the rate at which system components fail; or *time to failure*, that is, the time interval between beginning of system operation and occurrence of the first fault.

The time interval over which reliability is evaluated is generally domain dependent, and is chosen on the basis of the operational conditions under which the system is expected to operate. Notice that other factors, such as the possibility of periodic maintenance, may influence the way reliability is evaluated. For instance, in the avionic domain, periodic maintenance is scheduled after an appropriate length of service; therefore, it might be reasonable to evaluate reliability over the time interval between successive maintenance tasks. In other domains, maintenance may be impossible or inconvenient (e.g., for telecommunication satellites).

Finally, we note the difference between safety and reliability. A safe system is not necessarily reliable. For instance, a railway line in which all signals are set to red is intrinsically safe, but has null reliability. A safe system configuration such as the one just described is called a **failsafe state**. The existence of failsafe states is clearly important for safety reasons (e.g., in the railway domain, in emergency situations it is always possible to switch all signals to red, forcing all trains to halt and reaching a safe configuration). However, some domains do not possess any failsafe state (e.g., an aircraft during flight does not have any such state).

2.2.3 Availability

Availability evaluates the probability of a system to operate correctly at a specific point in time. The notion of availability is related to the one of reliability, but intrinsically different. In particular, while reliability evaluates the continuity of correct service for a given time interval, availability evaluates the correctness of service at a specific point in time. Alternatively, availability can be seen as measuring the percentage of time the system is providing correct service over a given time interval.

Clearly, it is domain dependent whether reliability or availability should be considered more important when evaluating system operation. In domains where the occurrence of a failure does not have catastrophic consequences, availability may be the primary concern. For instance, in telecommunication services, it may be reasonable to measure the effectiveness of service in terms of availability. In such domains, occurrence of faults may not be a problem, inasmuch as the service can be restored quickly (e.g., by shutting down and restarting a computer). On the other hand, in domains where a failure can have serious consequences or where maintenance is impossible, such as for satellites, reliability is more important than availability. Finally, availability is of critical importance in protection systems (e.g., shutdown system for nuclear power plants) that are expected to be used at specific points in time, rather than continuously.

Clearly, availability is different from safety for the same reasons that reliability is. A system brought into a failsafe state is neither reliable nor available, but is likely to be very safe.

2.2.4 Integrity

The notion of **integrity** refers to the possibility that a system will detect faults during operation. That is, integrity focuses on fault detection rather than fault tolerance (compare Section 2.8). The notion of integrity is linked to the notion of **fault recovery**, that is, the characteristic of a system being able to take appropriate measures to restore correct and safe operation after a fault has been detected. For some applications, recovery may be automatic, whereas for others it may require human intervention. In the latter case, upon detection of a fault, the system typically draws human attention to the event and relies on human operation for recovery. A related notion is the concept of **data integrity**, which is relevant in application fields where consistency of the data is of utmost importance. Data integrity refers to the possibility that the system will detect faults involving its internal database, and possibly recover by correcting the errors.

In the context of safety, the term **safety integrity** refers to the capability of a system to perform in a satisfactory way all the required safety functions, in the stated operational conditions and within a stated period of time. In the same context, the notion of **safety integrity level** is related to system classification in risk analysis, and is discussed in more detail in Section 3.3.

Finally, when used in particular expressions, such as **high-integrity systems**, the notion of integrity has a more general connotation, which involves notions of reliability as well as availability. In this view, a high-integrity system is the same as a dependable system, as described in Section 2.2.6.

2.2.5 Maintainability

Maintainability can be described as the possibility that a given system will be maintained. This includes both preventive actions that aim to avoid faults before they occur (e.g., periodic maintenance of an aircraft), and actions that can be taken to return a system that has failed to its operational conditions. Quantitatively, maintainability can be evaluated in terms of the so-called *mean time to repair*, that is, the average time to restore the system operation in case of a failure.

Maintainability is clearly domain dependent. In some domains, such as in avionics, routine maintenance is periodically scheduled, whereas in others (e.g., satellite service) it is impossible or simply too expensive. In some domains, maintenance can sometimes be performed during service by temporarily interrupting system operation, whereas in other domains maintenance can be performed only while not in service (e.g., in avionics, maintenance cannot be performed during flight).

2.2.6 Dependability

We conclude this section with the notion of **dependability**. A dependable system can be defined as a system for which reliance can justifiably be placed on the service it delivers (Laprie, 1991). In a broad sense, dependability includes most of the factors previously discussed, namely, safety, reliability, and availability, as well as security. According to Leveson (1995), integrating such different concepts into a combined notion can be misleading. For this reason, dependability is typically evaluated only qualitatively.

2.3 Classification of Faults

Faults can be categorized in several ways, for instance, depending on the equipment affected, their origin, the way they manifest themselves, and their duration. More specifically, we can classify faults depending on the following attributes:

- **Nature.** Depending on their nature, we can distinguish faults as being **systematic** or **random**. A systematic fault is inherent in a given system; for instance, it may be due to a wrong design or to a software bug. A systematic fault may or may not manifest itself. For example, a bug in a piece of code may become apparent only when the code is executed; nonetheless, the fault is always present and will not disappear unless it is explicitly removed. On the other hand, a random fault may occur nondeterministically at any time. Hardware components are typically subject to random faults. Hardware component faults may be due to aging and wear-out, or to stress conditions (i.e., operation of a component under environmental conditions for which it was not designed to work). Depending on its nature, it is also possible to classify faults as being **accidental** or **intentional**. The latter category includes faults due to malicious access or intrusion, and is typically related to security.
- **Equipment.** Depending on the equipment or part of the system that is affected, we can broadly distinguish between **hardware faults** and **software faults**. A software fault is an example of systematic fault, according

to the previous classification, whereas a hardware fault can be either random or systematic. A systematic fault of a hardware component can be due, for instance, to a wrong design or a wrong interconnection with other components.

■ **Phase of creation.** Depending on the phase in which faults are introduced, we can speak about **design faults** or **operational faults**. A design fault (for both hardware and software components) is inherent in the design of a given system, and it can be due either to a wrong specification or to a wrong implementation of the specification. An operational fault, on the other hand, occurs during operation of a given system; for instance, it may be due to normal wear-out of a hardware component, or caused by incorrect operation (e.g., a component can fail because of increased stress due to its operation under conditions that were not foreseen at design time).

■ **Extent and system boundaries.** Depending on the extent, we can classify faults as being **localized** (i.e., affecting only one system module) or **global** (i.e., affecting the system as a whole). Depending on the system boundaries, faults can also be categorized as **internal** or **external** with respect to the given system.

■ **Duration.** Depending on the duration, a fault can be **permanent** or **transient** (**sporadic**). A permanent fault is such that it does not disappear by itself; for instance, all systematic faults are permanent. On the other hand, random hardware failures may be transient; for example, it is possible for a sensor to produce a wrong reading due to transient conditions of the environment in which it is operating (e.g., due to electromagnetic radiation). Note that although a fault is transient, its effects may be permanent; that is, its effects may persist after the fault has disappeared. A further typology of faults is that of **intermittent** faults, that is, faults that may appear and disappear over time, due to environmental conditions or hardware characteristics (e.g., poor contact between two components in a circuit). Note that this classification refers to the faults themselves, and not to their effects. For instance, a software fault may externally appear as being intermittent, when it manifests itself only occasionally (e.g., due to timing considerations or race conditions), although it is in fact a permanent fault.

2.4 Fault Models

As seen in Section 2.3, faults can be classified in a number of ways. However, regardless of their classification, it is typically useful to provide a logical characterization of faults in terms of their effects on a given system. This kind of characterization can be formalized in terms of **fault models**. In other words, the definition of a fault model abstracts away concepts such as the nature and origin of faults, and concentrates exclusively on the effects at system level. Sometimes, timing aspects of the fault behavior may be considered, as well as their logical characterization.

Later in this section we present a few examples of fault models. Fault models may not be perfect representations of the effects of faults, but they can be extremely useful in engineering for a number of reasons (e.g., in testing and diagnosis). Fault models can be used to automatically generate test vectors for use during system testing or simulation (compare Section 5.7.1). By analyzing the impact of faults on the outputs of a given system, with the aid of a fault model, it is possible to generate specific test vectors that stimulate some erroneous system behavior. Given that exhaustive testing is typically out of reach, fault models can be used to reduce the number of test vectors to be tried, and improve testing coverage. Finally, fault models are useful in designing fault-tolerant systems, in that they help engineers understand the possible faults that might arise in a system, and estimate which of them should be tolerated. Fault tolerance is discussed in detail in Section 2.8.

Although in principle it is possible to define fault models for both hardware and software components, hardware fault models are easier to identify and formalize than software fault models. This is due to the inherent complexity of foreseeing the effects of software faults (e.g., the effect of a wrong calculation, the effect of an array overflow or of the use of uninitialized variables). Traditionally, the following hardware fault models have been considered: the (single or multiple) *stuck-at* fault model, the *stuck-open* (and *stuck-closed*) fault models, the *bridging* fault model, and the *delay* fault model. These fault models are described in more detail in the remainder of this section. Some of these models, and additional ones, are supported by the FSAP platform (FSAP, 2009; Bozzano and Villafiorita, 2007) (see Appendix B for more details). Although these fault models do not exhaust the types of erroneous behavior that can occur, in practice a large number of faults may be conveniently and adequately represented in this way.

A further classification distinguishes **transistor-level** (also called **atomic-level**) faults from **gate-level** faults. The former category models faults of elementary constituents of circuits, such as transistors, resistors, and capacitors. Typical faults can be described in terms of open circuits or broken elements, and wrong interconnections such as short circuits or bridges. The latter category of faults abstracts away from these low-level details and focuses on the logical behavior of gates (e.g., logical AND, logical OR). Although transistor-level fault models may represent a good model in terms of explanation of the mechanisms underlying a number of faults, their use is limited due to their inherently higher complexity when analyzing digital circuits.

2.4.1 The Stuck-At Fault Model

The **single stuck-at** fault model describes the effect of a fault in a digital circuit in terms of its visible behavior. In particular, it assumes that a module (e.g., a logical gate) will behave as if one of its inputs or outputs is stuck at a particular value (logical

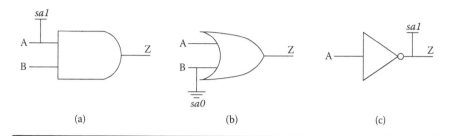

(a) (b) (c)

Figure 2.1 Three examples of stuck-at faults: (a) an AND gate with a stuck-at-1 fault on input A, (b) an OR gate with a stuck-at-0 fault on input B, and (c) a NOT gate with a stuck-at-1 fault on output Z.

0 or logical 1, for logical gates). When inputs are involved, the function associated with the module or gate is assumed to be unchanged by the fault (and computed using the faulty inputs); whereas for an output, the model prescribes that it is stuck at the given value independently of the inputs.

A stuck-at fault at logical 0 is also called *stuck-at-0* (shortened to *sa0*), whereas a stuck-at fault at logical 1 is called *stuck-at-1* (shortened to *sa1*). We exemplify this on an AND, an OR, and a NOT logical gates in Figure 2.1. The corresponding truth table of these faulty gates are shown in Figure 2.2.

Typically, it is assumed that a stuck-at fault is permanent. Furthermore, to simplify the analyses, it is often assumed that only one fault for each line can be active at any time (single stuck-at fault model), although nothing prevents the consideration of multiple faults. Using multiple faults increases the percentage of defects that can be covered by the model. The single stuck-at model is one of the most successful and pervasively used fault models. Empirical studies indicate that about 90% of the defects in CMOS (complementary metal-oxide-semiconductor) circuits can be described in this way. Often, more complex faults, such as the ones described in Sections 2.4.2 and 2.4.3, can be mapped to sequences of single stuck-at faults.

A	B	Z
0	0	0
0	1	1
1	0	0
1	1	1

A	B	Z
0	0	0
0	1	0
1	0	1
1	1	1

A	Z
0	1
1	1

(a) (b) (c)

Figure 2.2 The corresponding truth tables for the faulty gates in Figure 2.1.

2.4.2 The Stuck-Open and Stuck-Closed Fault Models

The **stuck-open** (also called **stuck-off**) fault model is an example of a transistor-level fault model and is used to describe erroneous conditions associated with CMOS gates, which cannot always be described in terms of stuck-at faults. The low-level explanation of the erroneous behavior may be related to an internal open-circuit or short-circuit that causes the output of a gate to assume a value that does not correspond either to logical 0 or logical 1. A similar model is the so-called **stuck-closed** (also called **stuck-on**) fault model.

The effect of such faults is highly dependent on the exact circuit configuration. Furthermore, additional and subtle effects may arise, in that some gates may present a "memory effect" that causes the output to maintain the value it had in its previous state. Such effects may turn a combinational circuit into a sequential one, and may originate very complex behaviors.

2.4.3 The Bridging Fault Model

The **bridging** fault model describes situations in which two circuit components are shorted together (e.g., two wires in a circuit). The junction between the elements is typically assumed to have low resistance (as a consequence, in logical terms the elements joined together assume the same value). In principle, this model can be applied both to low-level components (e.g., transistors) and to logical gates (with or without feedback); hence both transistor-level and gate-level faults can be described in this way. Typically, this kind of fault is considered permanent.

A bridging fault model cannot be always represented using the single stuck-at model. Traditionally, two effects of bridging faults are considered: **wired-AND** (also called **positive logic bridging**) and **wired-OR** (**negative logic bridging**). In the former case, the effect of the bridge is equivalent to replacing the piece of faulty circuit with the output of an AND gate, whose inputs are the components shorted together. Similarly, in the latter case, the faulty circuit can be replaced with an OR gate. This is exemplified in Figure 2.3. When multiple bridges within the same circuit are involved, or when bridges involve feedback paths, very complex behaviors may arise, which can result in turning a combinational circuit into a sequential one, or producing oscillations in the circuit.

2.4.4 The Delay Fault Model

The delay fault model typically describes situations in which a circuit component eventually assumes the expected value, but with a certain delay (more rarely, the value can be assumed more quickly). This kind of effect can be explained in terms of the physical defects of the involved components at transistor level.

Traditionally, two alternative models are considered: **gate delay** (also called **transitional**) fault model, in which a delay of a single circuit line or gate is considered,

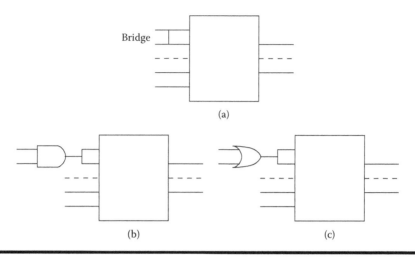

Figure 2.3 Two examples of bridging faults: (a) a generic module with a bridging fault on two of its inputs; (b) the logical interpretation of the bridging fault in positive logic; and (c) the logical interpretation of the bridging fault in negative logic.

and the **path delay** fault model, which considers the total delay along a path from some inputs to some outputs. The second model accounts for compensations that may arise as a consequence of the different speeds of different circuit elements.

2.5 Managing Faults

The design and safety assessment of safety-critical systems is closely related to the management of faults. There are several ways to deal with faults. Traditionally, the following techniques can be distinguished (Storey, 1996).

- **Fault avoidance.** Tries to keep faults out of the design, for instance, using formal methodologies and quality control to ensure, to the wider extent possible, that a given system is fault-free (e.g., that a piece of software is bug-free). Use of formal methods for system specification and design, which is the object of Chapter 5, has fault avoidance as one of its main objectives.
- **Fault removal.** Aims at removing faults from the products once they have been entered as a consequence of improper design or wrong implementation. Traditionally, fault removal can be carried out by performing system testing (e.g., of a piece of software) and removing all the bugs that have been discovered. Alternatively, formal verification techniques can be used (compare Chapter 5).
- **Fault detection.** Used to detect faults of a product during service, so that adequate countermeasures can be taken to prevent the fault from manifesting

itself as an error or failure. There are several techniques that can be used for detecting faults; we describe some of them in more detail in Section 2.6.

■ **Fault tolerance.** Used to allow a given system to operate correctly in the presence of faults; for instance, it can be achieved by employing a redundant architecture. Fault tolerance can be combined with fault detection; in fact, many fault-tolerant architectures use fault detection techniques to recognize the occurrence of faults and take appropriate countermeasures (e.g., a component that has been detected as faulty can be disabled and removed from the redundant arrangement).

Typically, a combination of these techniques is used to ensure safe operation of a safety-critical system. In the remainder of this chapter we focus on fault detection and fault tolerance techniques; fault avoidance and fault removal are the subject of Chapter 4 and Chapter 5.

2.6 Fault Detection

Fault detection techniques (Storey, 1996) are used to diagnose the presence of faults, with the intent of preventing the fault from causing an error or, even worse, a failure. Fault detection can be coupled with fault tolerance, and with fault recovery strategies, whose goal is to maintain or restore correct or safe operation. In general, fault detection techniques are complicated by the partial observability of the system under analysis.

Fault detection techniques can be divided into three broad categories, as follows:

1. *Model-free approaches* are typically based on data analysis. They attempt to detect faults by comparing actual data with historical data collected both during nominal and faulty system operation. This mechanism requires the prior acquisition of statistical data.
2. *Knowledge-based approaches* exploit expert knowledge to detect faults. Fault detection systems in this category may assume the form of expert systems using knowledge-based (possibly heuristic) rules that detect a malfunction from a set of symptoms. Another example is given by approaches based on neural networks that combine heuristic knowledge with data. This approach requires the prior acquisition of knowledge.
3. *Model-based approaches* are based on a description of the nominal system behavior in terms of a nominal model. The behavior of the actual system is compared with the behavior expected from the nominal model, and any deviation is taken as an indication of malfunction.

Regardless of the approach, fault detection techniques can be distinguished, depending on whether they are performed **online**, that is, during system operation, possibly with real-time constraints, or **offline**, for instance, for *post-mortem diagnosis*.

Moreover, fault detection techniques can be used for detecting either software or hardware faults, and they themselves can be implemented either in software or in hardware.

2.6.1 Functionality Checking

We group under this category several techniques that detect the wrong operation of hardware components using routines to check their functionality. Typical examples are routines to test and check memory and processor operations, and network communication. A typical example is given by routines that check memory functionality by periodically issuing writing and reading operations to selected locations and checking the correctness of the produced results. In the case of read-only memories (ROM), the checks can be performed by computing checksums and comparing them with expected results, which are typically stored in ROM. A measure of coverage must be considered in the design of such routines. Similar routines that perform a sequence of operations, whose results can be checked against an expected result, can be used to check correct operation of a processor. Finally, messages can be sent periodically over a communication network to ensure that communication links are working as expected.

2.6.2 Consistency Checking

Techniques in this category detect faults by comparing the actual outputs of software or hardware components with an expected range of values. A different example is given by routines that check the consistency of sets of data, for instance, the consistency of the file system (in some cases offering the possibility to fix the errors encountered).

2.6.3 Signal Comparison

Signal comparison consists of checking different signals within a redundant system that are assumed equal, for instance, the same signal in replicated components. A special case of signal comparison is given by *checking pairs*, which detect differences in signals produced by the same identical modules. If two modules produce the same signal, they are assumed to be working correctly.

2.6.4 Instruction and Bus Monitoring

There are several ways to check correct operation of a processor. Typically, the operation code and operand of each instruction is checked, after being fetched from memory and before being executed, for possible corruption. If the operation is illegal, typically an exception can be raised and execution suspended, allowing possible recovery actions from the fault. The bus can also be monitored for illegal access to memory locations that are allowed for a given process. Any illegal access is reported as an error to the processor.

2.6.5 Information Redundancy

We group in this category several techniques that use forms of information redundancy to detect faulty data. These techniques include parity checks, computation of checksums, and several forms of error correcting codes. These techniques are typically used for detecting errors (e.g., communication errors over unreliable and noisy communication links) and mitigating their effect. Error detection codes require retransmission of faulty data, whereas error correcting codes allow, up to a certain degree, automatic correction of corrupted data without the need for retransmission. Error correcting codes are typically based on sending additional check bits, together with the original data, which are used for error detection and correction. More complex techniques, such as cyclic redundancy checks, are based on the properties of finite fields and polynomials over such fields.

2.6.6 Loopback Testing

Loopback testing techniques aim to verify that a signal departing from a given location reaches its destination unchanged. This type of check is carried out to detect faults that are associated with shorted parts of circuits or bridges (compare Section 2.4.3). These techniques require the presence of an independent return path for the signal to be tested to its source. Loopback testing can also be used to test communication lines over a network, or correctness of data sent over a bus. In the case of communication networks, the loopback can be designed in such a way to test a plurality of network connections from a source to a destination. Suitable protocols are used to check the correctness of loopback messages.

2.6.7 Watchdog and Health Monitoring

A watchdog timer is a counter that is initialized with a given value, greater than zero, and is decremented at regular time intervals. Periodically, the processor is required to reload the initial value of the counter. If, for any reason (e.g., due to a system crash), the value of the counter is not reloaded, its value eventually reaches zero, which signals that an error has occurred. Typically, the response to such an event can be an automatic system reset. Techniques for health monitoring are used to test the health of a unit being monitored by periodically sending messages to it. A fault is detected whenever one or more messages are not acknowledged by the given unit.

2.7 Fault Prediction

Fault prediction is concerned with evaluating the likelihood that a given system or component will fail over a period of time. Fault prediction can be used for a number of reasons. Evaluating the probability of failure of basic components is needed to estimate the reliability of the whole system, which, in turn, is a necessary step during system design and for certification purposes. For instance, different fault-tolerant architectures can be evaluated and compared with respect to their reliability.

The evaluation of the fault probability is highly dependent on the nature of the fault itself. In the case of systematic faults (e.g., a design fault), it is extremely difficult to measure the likelihood that they will occur. On the other hand, for random faults (e.g., hardware component faults), it is possible to estimate the probability of a fault on the basis of past experience and data. While it is impossible to predict the exact moment in time when a given device will fail, statistical data collected in the past can be used to estimate failure rates. This estimation, together with the use of appropriate fault models, can be used, in turn, to design system strategies to deal with these faults, for instance, the use of fault-tolerant architectures and fault detection strategies. Care must be taken in evaluating the fault probability of a component under the environmental conditions in which the component is expected to operate. If a component is subject to stress conditions, due to its operation under non-nominal conditions, then its failure rate may be significantly higher.

A more controversial debate concerns the evaluation of the failure rate for software components. A software fault is, by definition, a systematic fault, and, as such it is meaningless to estimate the probability of occurrence of such faults using statistical analysis. A software fault will occur every time the faulty piece of code is executed. A different line of thought, however, is followed by engineers that claim that unidentified software faults, for practical analysis purposes, can be considered as if they were randomly distributed within the code. Hence, for these engineers, statistical analysis can be used to predict the occurrence of software faults.

The use of statistical data for evaluating failure rates is the basis for reliability modeling, which is discussed in more detail in Section 2.10.

2.8 Fault Tolerance

As mentioned earlier, fault tolerance (or graceful degradation) is the capacity of a system to operate properly on the hypothesis of the failure of one (or more) of its components. Fault tolerance is an essential or desired requirement for a wide range of systems that do not necessarily have to be safety critical. The most well-known example of a redundant system is, in fact, the Internet because it was designed to

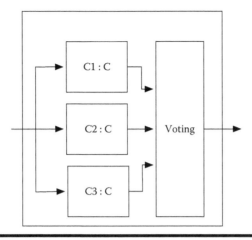

Figure 2.4   Triple modular redundancy.

operate even if some of its components did not work[1]. Fault tolerance is achieved through some form of replication of the critical components. Different schema exist that allow for different levels of fault tolerance (to match, in turn, different levels of criticality in case of failure) and other constraints, such as space, costs, power consumption, etc. In the remainder of this section we look at some of the most common ones.

2.8.1 Triple Modular Redundancy

Possibly the most well-known architecture for achieving fault redundancy is triple modular redundancy (TMR), first proposed by von Neumann (1956) and by Shannon and Moore (1956), although their work focused on building reliable logic structures from less reliable components.

The TMR architecture is based on replication of the critical component, as illustrated in Figure 2.4. In the figure, the component performing a critical function (*C*) is replicated three times (*C1*, *C2*, and *C3*). A voting module is responsible for taking the inputs of the replicated modules and deciding which output must be propagated.

Different policies can be implemented by the voting component. The most common voting mechanism is based on majority voting. The output produced by at least two components is the one propagated to the other elements of the architecture. Other policies are, however, possible: We mention the possibility of choosing the average or the median value produced by the modules.

---

[1] Fault tolerance is achieved with the packet-switching protocol. See (Various Authors, 2009b; Griffin, 2009) for a history of the Internet and a discussion about packet switching and resilience of the Internet to a nuclear attack.

The TMR architecture provides fault tolerance under the assumption that the following main hypotheses hold:

1. **At most, one replicated component fails.** This is obvious for TMR with majority voting because the voting system needs at least two equal inputs to propagate the information. This, however, also holds for other implementations of the voting mechanism (such as average and median): In fact, although the system might still produce the correct output even if two of the components fail (because the median or the average could correspond to the correct output), it would do so by chance.

2. **The voting mechanism does not fail.** The voting component is not replicated, and any failure of the voting component could result in the wrong data being propagated to the next elements in the architecture, independent of whether the redundant modules are working.

3. **There are no systematic failures.** That is, the different instances of the replicated component will fail in statistically independent ways. The hypothesis holds for causes of failures due to, for example, wear-out of a component. However, problems in the requirements, design, and production of the redundant components might violate the independence hypothesis and thus make the TMR architecture useless.

4. **Isolation of the failed component is not an issue.** The TMR works under the assumption that masking a faulty output is sufficient to guarantee that the system will operate as expected. This, however, might not always be the case. Think, for instance, of situations in which the failed component interferes with the behavior of the other working components.

Different variations on the TMR architecture have therefore been proposed to overcome one or more of the issues mentioned above. We present them in the next few sections.

2.8.2 Dealing with Multiple Failures

Tolerance to multiple failures of the critical component can be achieved by increasing the redundancy. That is, rather than using three replications of the same component, we use N replications of the component, where N is an odd number, if majority voting is used to determine the output. This architecture is called **N-modular redundancy**. The N-modular redundancy tolerates up to $(N-1)/2$ failures.

Figure 2.5 show an example of a N-modular redundant architecture, with $N = 5$. The system tolerates up to two failures $((5-1)/2)$.

Notice, however, that practical considerations, such as costs, power consumptions, physical constraints, etc. might make the deployment of N-modular redundant architecture unfeasible.

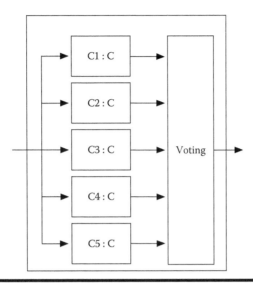

Figure 2.5 *N*-modular redundancy.

2.8.3 Dealing with Failures of the Voting Component

The voting component of a TMR performs a simple operation. In principle, therefore, it can be implemented with a simple architecture whose reliability might be simpler to achieve than for the replicated components. Storey (1996), for instance, shows an implementation of a majority voting system requiring four *nand* gates for every bit in output from the replicated components.

The voting component, however, is a single point of failure of the TMR architecture; and when the reliability of the voting component cannot be proven or is not acceptable, other solutions must be adopted. One of such approaches consists of replicating the voting component. It is illustrated in Figure 2.6, where the voting module is replicated three times.

The architecture is now tolerant to one failure in one the three voting modules and one failure in one of the replicated modules, because two of the three outputs of the module are correctly propagated to the next element in the architecture. Notice, however, that to accommodate the different number of outputs (three instead of one), this architecture can be used only in a series of cascading TMRs.

A TMR with triple voting architecture was used for the LVDC (launch vehicle digital controller) of the Saturn V rockets. The LVDC consists of three identical sets, or channels, of logic. Each channel, in turn, is composed of seven stages. Voting circuits at the end of each stage mask wrong outputs through majority voting, as illustrated in Figure 2.7 (taken from IBM (1964)). See IBM (1964) for more details.

Variations in the architecture illustrated in Figure 2.7 have been proposed. For instance, Lee and Lee (2001), and Lee et al. (2007) propose an architecture in which only one voter is used at the end of each stage of the cascading TMRs. The

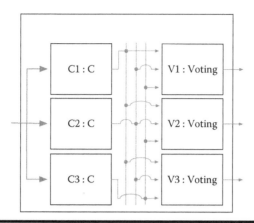

Figure 2.6   Triple modular redundancy with triple voting.

reduced number of voters imposes a constraint on the number of coincident failures of the voting components. The architecture is demonstrated to be robust to multiple failures of the voting components, as long as they are at least three stages apart. See Lee and Lee (2001), and Lee et al. (2007) for more details.

2.8.4 Dealing with Systematic Failures

The TMR architecture is not robust to systematic failures, that is, errors in the specifications, in the design, or in the production of the replicated components, or of their sub-components. In case of systematic errors, in fact, all replicated components will fail in exactly the same way. The voting component, when the error occurs, will receive the same wrong output from all the replicated modules and will not be able to mask the failure. Systematic errors are particularly relevant for software because all software failures are due to specification or design errors.

Systematic faults (also known as **common mode faults**) are mitigated by differentiating the components used to implement the critical function. For instance,

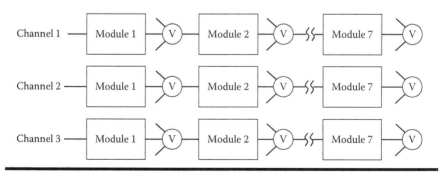

Figure 2.7   Saturn V launch vehicle digital computer.

three different microprocessors could be used to implement the redundant processors of a TMR. Protection from systematic failures in software is usually achieved through the independent development of different versions of the same system. The approach is called *N*-version programming and is described in more detail in Chapter 4, together with some of the critiques that have been raised to the approach. Here we just mention that the approach requires independent teams and might be reinforced through the use of different programming languages, different compilers, and different target architectures. The main drawback to this approach are increased development and maintenance costs. Finally, we mention that even when independent implementations are provided, errors in the requirements of the component might still constitute a single point of failure.

2.8.5 Fault Tolerance and Fault Detection

The last hypothesis we mention about the TMR architecture is that the faulty component does not interfere with the rest of the architecture and isolation, therefore, is not necessary. That is, ensuring that the errors are not propagated is a sufficient strategy for dealing with failed components. However, this might not always be the case. In such situations, fault tolerance requires some form of detection of the failure and adaptation of the system, so that when a failure is detected, the faulty component can be properly isolated.

The general architecture is illustrated in Figure 2.8. A fault detector monitors the behavior of the critical component to detect failures. If a failure is detected, the fault detector switches the control to the redundant component, also called *stand-by component* and possibly switches off the failed component.

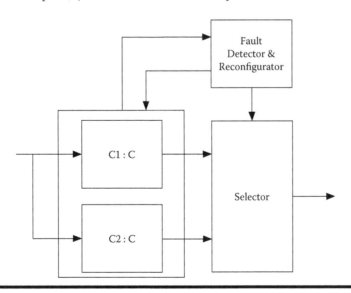

Figure 2.8   Active redundancy.

The architecture presented above can be further refined according to the behavior of the stand-by component:

- In the **hot stand-by** architecture, the redundant component runs in parallel to the standard component. Switching control to the stand-by component is thus immediate. The drawback of this arrangement is that the stand-by component is subject to the same operational stresses and wear-out as the standard component.
- In the **cold stand-by** architecture, the redundant component is turned off by default and is switched on when the first component fails. The architecture clearly allows for reduced stress on the stand-by component and minor power consumption, to name two advantages. However, the architecture assumes that the start-up time of the standby is faster than the time required to provide the next output. Other assumptions are that the component cannot fail if not powered, and that the component will actually start when required to do so.
- Finally, in the **warm stand-by** arrangement, the redundant component is actually powered, but idle. Warm stand-by is often used for mission-critical IT systems in which the standby system is kept in sync with the primary system and used if a failure occurs in the primary system.

One important aspect of the stand-by architecture is how faults are detected. Practical solutions are often based on the replication of the critical module. The output produced by one module is compared with that produced by an identical spare; if the outputs differ, the critical module is supposed to have failed (although, in principle, the redundant copy might have failed).

2.8.6 Dealing with Transient Failures

As pointed out earlier, the TMR is not robust to multiple failures. It is thus vital to be able to detect and replace failed components during regular maintenance of the system, to ensure that all components operate as expected. However, owing to their very nature, transient failures might be difficult to reveal during maintenance.

In such situations, the TMR architecture can be extended to include a fault detection system to log failures.

2.8.7 Emergency Shutdown Systems

If a safety-critical system is failsafe—namely, if shutting the system down brings it to a safe state—an architecture often used is the so called **emergency shutdown architecture**, which we show in Figure 2.9.

The architecture is composed of a critical component, a shutdown system, and, possibly, a system to diagnose the correct functioning of the shutdown system. The

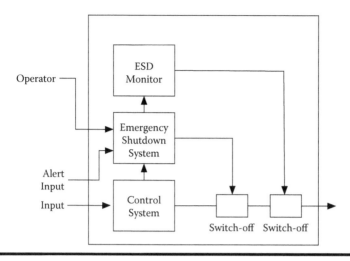

Figure 2.9   Emergency shutdown architecture.

shutdown system might take the same inputs of the critical component, have special controls for manual initiation of the shutdown procedure, or monitor the behavior of the critical component. Duplication of input channels is usually adopted to ensure that inputs are not a single point of failure or to measure different values (e.g., the critical component might read the current level of a chemical in a tank and the emergency shutdown system read the value of a "close to maximum level" sensor). If the shutdown system detects a dangerous situation or is required to do so by a user command, it disconnects the control system. The diagnosis system is responsible for initiating the shutdown if it recognizes an anomaly in the shutdown system itself.

Nuclear reactors have emergency shutdown systems, although their architecture differs a bit from the one we have shown because they directly act on the core to stop the nuclear reaction, rather than switching off the control system. One system used by various types of plants is based on the insertion of graphite rods in the nuclear core, as graphite absorbs neutrons and thus reduces fission of the nuclear material. The shutdown system is thus responsible for commanding the insertion of the rods if required by the operator. In the case of the Chernobyl accident, unfortunately, the emergency shutdown procedure failed when the rods got stuck before completely entering the core. Failure of the shutdown system, however, is just one of the events that led to the accident. See Various Authors (1987) for the official NUREG report, and Various Authors (2009a), and Gonyeau (2009) for some very good starting points on the Internet.

2.8.8 An Example of Redundant Architecture: Fly-by-Wire Systems

Traditionally, fault tolerance in aircraft was achieved through the use of independent and diverse systems for actuating pilots' commands. For instance, the control system of the airplane surfaces could be based on a redundant set of channels based on different technologies: electric wiring, hydraulic commands, and mechanical commands. Over the years, digital computers have proven a more effective solution, and the so-called fly-by-wire (FBW) systems have been gradually introduced into military and commercial aircraft. We conclude this overview of fault-tolerant architectures by presenting the generic architecture of an FBW system because it allows us to demonstrate most (if not all) of the concepts described in this section. It is shown in Figure 2.10.

The FBW system is composed of two redundant systems: left and right. Both the left and right systems might be implemented with a redundant architecture, such as, for instance, a TMR. The left and right systems take inputs from the cockpit (left pilot and right pilot in the diagram) and from redundant sensors, and provide outputs to the autopilot, which is responsible for delivering the commands to control the airplane surfaces. Only one pilot interface is active at a time (the active interface can be switched by the pilots), so that both systems receive inputs only from one side of the cockpit.

The FBW system can have two operating modes: independent and dependent. In the dependent operating mode, only one side (e.g., either the "left" or the "right" system) is active and the other works as a hot spare. If the LHS and RHS are

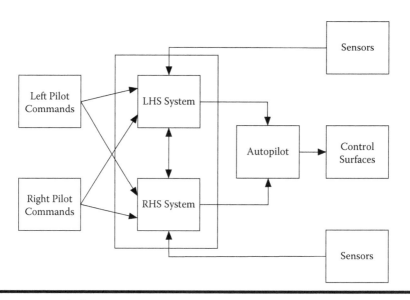

Figure 2.10   Fly-by-wire System.

implemented with a TMR architecture, single failures of the active side can be detected and cause the switch to the hot spare.

In the independent mode, both systems operate independently and provide input to the autopilot. The autopilot, in turn, implements a voting policy to verify whether the two systems agree on the command and is responsible of signaling a problem in case of disagreement. The command the autopilot delivers to the actuators is the average of the values produced by the "left" and the "right" systems. However, if the difference between the values of the two systems is above a given threshold, the autopilot signals a problem and automatic controls are disconnected.

In actual implementations, to decrease the risk of common failures, the implementation of the "left" and "right" systems might be based on different hardware and software architectures and electrical power provided by independent systems (e.g., the left and the right engine).

See Siewiorek and Narasimhan (2005) for a more detailed description and an overview of redundant architectures for airplane and spacecraft, and see Miller and Whalen (2005) for the verification and validation of an FBW system using model checkers.

2.9 Fault Coverage

We have seen in the introduction of this chapter that faults can be dealt with in different ways, namely, by means of fault avoidance, fault removal, fault detection, and fault tolerance. The concept of **fault coverage** applies to each of these different techniques, and is a measure of their degree of success. More practically, fault (avoidance, removal, detection, or tolerance) coverage can be measured by evaluating the percentage of faults that can be, respectively, avoided, removed, detected, or tolerated.

In general, there is no simple and reliable way to precisely evaluate fault coverage. In the case of fault avoidance, it is very difficult to estimate how many faults have been prevented from entering a design as a consequence of a specific design methodology or technique. In some cases, this estimation can be done indirectly by evaluating the quality of the resulting product with respect to that of similar products that were developed using alternative methodologies (compare, for instance, the measures given in Section 5.9 for some industrial-oriented applications of formal methods). In the case of fault removal, detection, and tolerance, fault coverage could in principle be experimentally estimated through testing. However, for real systems, it is practically impossible to test all possible faults and occurrences therefrom. Moreover, the estimation of fault coverage requires some assumptions on the type of faults that can occur and that should be considered in the evaluation.

In practice, fault coverage is evaluated on the basis of some specific fault model; candidate fault models are the ones described in Section 2.4; in particular, the single-stuck-at fault model (compare Section 2.4.1) is one of the most used models. Some assumption on the occurrence of faults is typically made to make the analysis feasible. For instance, the analysis may be limited to permanent faults, and constraints may be put on the number of faults that can exist in a design at any given time (e.g., single fault). These assumptions are due to practical considerations. It can be shown that in a system with N fault sites (e.g., inputs or outputs of gates in a circuit), there are $2N$ single stuck-at faults, and $3^N - 1$ different multiple faults. In the case of the bridging fault model (compare Section 2.4.3), the situation is even worse. In fact, there are $\binom{N}{M}$ possible bridging faults connecting M out of the total N nodes, which corresponds to the order of N^2 potential faults between two nodes, and the order of N^3 faults between three nodes. For these reasons, having complete fault coverage is out of reach in most practical situations. Nonetheless, fault coverage based on specific fault models, under some assumptions, is still a useful measure in practice.

2.10 Reliability Modeling

We conclude this chapter by discussing the basics of reliability theory (Rausand and Høyland, 2004). In particular, we introduce a few quantitative measures for system reliability and discuss their interrelationships. Following the discussion in Section 2.7, we focus the presentation on hardware random faults that can be analyzed with the aid of statistical data, and we do not deal with systematic faults. We first define reliability for individual components, and then in Section 2.11 we derive reliability measures for a complete system. Moreover, we make the assumption that individual components that fail cannot be repaired. As discussed in Section 2.2.2, reliability can be defined as the capability of a system to perform a required function correctly, under given operational conditions, for a given interval of time. Hence, from a quantitative point of view, reliability can be defined as a function of time, usually denoted $R(t)$. Informally, we define $R(t)$ as a probabilistic function, namely as the probability that the component will perform correctly for the time interval $(0, t]$. Conversely, the probability that the component will *not* perform correctly, that is, the probability that it will fail within the time interval $(0, t]$, is called **failure probability** (alternatively, **unreliability**) and usually denoted $Q(t)$. By definition, we have that

$$R(t) = 1 - Q(t) \tag{2.1}$$

If we denote with T the **time to failure**, that is, the point in time when the component fails for the first time, we have that

$$Q(t) = P(T \le t) \tag{2.2}$$

that is, the failure probability at time t is equivalent to the probability that the time to failure T of the component is less or equal than t. Similarly, we have that

$$R(t) = P(T > t) \tag{2.3}$$

In the following we assume that the time to failure is continuously distributed, with distribution function $f(t)$. In symbols,

$$Q(t) = P(T \le t) = \int_0^t f(u)\,du \tag{2.4}$$

where $t > 0$. The function $f(t)$ is called the **failure probability density**. It can be defined as

$$f(t) = \frac{d}{dt}Q(t) = lim_{\Delta t \to 0}\frac{Q(t + \Delta t) - Q(t)}{\Delta t} = lim_{\Delta t \to 0}\frac{P(t < T \le t + \Delta t)}{\Delta t} \tag{2.5}$$

Evidently, we also have that

$$f(t) = -\frac{d}{dt}R(t) \tag{2.6}$$

For a sufficiently small time interval Δt, we can approximate

$$P(t < T \le t + \Delta t) \approx f(t) \cdot \Delta t \tag{2.7}$$

A closely related notion is the **failure rate** of a component. Informally, the failure rate represents the frequency of failure of the component, that is, the number of failures of the component that occur within a given time interval. The failure rate is usually denoted $z(t)$ and defined on the basis of a conditional probability, as follows:

$$z(t) = lim_{\Delta t \to 0}\frac{P(t < T \le t + \Delta t \mid T > t)}{\Delta t} = lim_{\Delta t \to 0}\frac{Q(t + \Delta t) - Q(t)}{\Delta t}\frac{1}{R(t)} \tag{2.8}$$

The expression $P(t < T \le t + \Delta t \mid T > t)$ represents the conditional probability that the component fails in the time interval $(t, t + \Delta t)$, assuming that it has not failed in the time interval $(0, t]$.

From Equation (2.5) we have that

$$z(t) = \frac{f(t)}{R(t)} \tag{2.9}$$

For a sufficiently small time interval Δt, we can approximate

$$P(t < T \leq t + \Delta t | T > t) \approx z(t) \cdot \Delta t \tag{2.10}$$

Notice the difference between Equations (2.7) and (2.10). While the failure probability density function is related to the probability of a component failing in the time interval $t < T < t + \Delta t$, considered at time 0, the failure rate is related to the probability of a component failing in the same time interval, under the hypothesis that it has survived up to time point t.

The failure rate can also be used as a measure of the frequency of failure of a number of different components. In this case, the failure rate can be seen as the instantaneous rate at which the components are failing, with respect to the components that are still working correctly.

From Equations (2.6) and (2.9), we have that

$$z(t) = -\frac{d}{dt} R(t) \frac{1}{R(t)} = -\frac{d}{dt} \ln R(t) \tag{2.11}$$

Given that $R(0) = 1$, it follows that

$$\int_0^t z(t) dt = -\ln R(t) \tag{2.12}$$

hence

$$R(t) = e^{-\int_0^t z(u) du} \tag{2.13}$$

Equation (2.13) shows that the reliability function is uniquely determined by the failure rate $z(t)$. From Equations (2.9) and (2.13) we also have that the failure probability density is given by

$$f(t) = z(t) e^{-\int_0^t z(u) du} \tag{2.14}$$

Normally, the failure rate varies with time. Many electronic components show the behavior depicted in Figure 2.11, which is known, for obvious reasons, as a *bathtub curve*. Their lifetime can be divided into three distinct periods. At the beginning, components show high failure rates that can be explained as *infant mortality*, that is, the presence of defects that have survived the manufacturing phase. At the end of their life, components again show high failure rates that can be described as *wear-out*, that is, are due to aging. Finally, in the middle part of their lifetime, components show a relatively constant failure rate. This constant failure rate is typically denoted λ. For the specific case of constant failure rate, Equation (2.13) assumes the simpler form

$$R(t) = e^{-\lambda t} \tag{2.15}$$

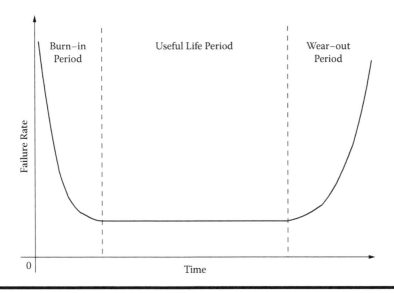

Figure 2.11   A bathtub curve.

This relationships between reliability and time is known as the *exponential failure law*, and the related failure rate distribution is called *exponential distribution* with parameter λ). The failure probability density in the case of an exponential distribution is given by

$$f(t) = \lambda e^{-\lambda t} \tag{2.16}$$

for $t > 0$. In practice, the failure rate for different components may show very different behaviors. Several distributions are discussed in Rausand and Høyland (2004).

We conclude this section by discussing a further reliability measure called the **mean time to failure** (MTTF), that is, the expected time a component is expected to operate before the first failure occurs. It is defined as follows:

$$MTTF = \int_0^\infty t f(t) \, dt \tag{2.17}$$

It can be shown that

$$MTTF = \int_0^\infty R(t) \, dt \tag{2.18}$$

In the case of an exponential distribution, the mean time to failure is given by

$$MTTF = \int_0^\infty e^{-\lambda t} dt = \frac{1}{\lambda} \qquad (2.19)$$

Hence, it is the inverse of the constant failure rate.

Finally, we discuss a quantitative measure for availability. Availability, as discussed in Section 2.2.3, can be evaluated as the percentage of time a given system or component is operational, measured over a given time interval. This can be expressed as

$$Availability = \frac{MTTF}{MTTF + MTTR} \qquad (2.20)$$

where $MTTR$ (**mean time to repair**) measures the average time needed to repair the system or component (including detection of the fault and possible system reconfigurations). The quantity $MTTF + MTTR$ is known as **mean time between failures** and denoted $MTBF$.

2.11 System Reliability

After discussing the reliability of individual components, in this section we look at techniques to estimate the reliability of complete systems. In particular, we look at two prototypical system structures, namely, the **series structure** and the **parallel structure**. Typically, the reliability of more complex layouts can be estimated by isolating their series and parallel sub-structures and combining their reliabilities using the techniques discussed below.

Throughout this section we assume that N components are given. We use R_i to denote the reliability function of component i, for $i = 1, \ldots, N$, and similarly for the functions Q_i, z_i, and the failure rates λ_i. We assume the same hypotheses as in Section 2.10, that is, we focus on hardware random faults, and we assume that components cannot be repaired. Moreover, we assume that components are independent, that is, a fault of one component does not affect the remaining components.

2.11.1 Series Structures

In a series structure, a system is composed of N independent sub-components, such that a failure of either of them will cause the complete system to fail. A series interconnection is schematically represented as in Figure 2.12.

Figure 2.12   A series structure.

In a series structure, the reliability function is the product of individual reliabilities, that is,

$$R(t) = \prod_{i=1}^{N} R_i(t) \tag{2.21}$$

From Equation (2.13) we have that

$$R(t) = \prod_{i=1}^{N} e^{-\int_0^t z_i(u)\,du} = e^{-\int_0^t \sum_{i=1}^{N} z_i(u)\,du} \tag{2.22}$$

The failure rate of a series structure is the sum of the failure rates of individual components:

$$z(t) = \sum_{i=1}^{N} z_i(t) \tag{2.23}$$

In the special case of constant failure rates (exponential distribution with parameters λ_i), the system failure rate is given by

$$\lambda = \sum_{i=1}^{N} \lambda_i \tag{2.24}$$

and the mean time to failure, by Equation (2.19), is given by

$$MTTF = \frac{1}{\sum_{i=1}^{N} \lambda_i} \tag{2.25}$$

Note that the failure rate of a series structure is constant if the failure rates of individual components are constant.

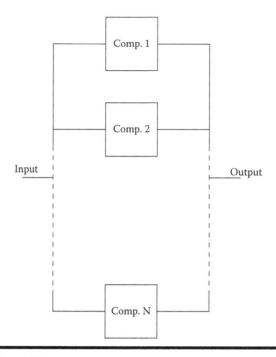

Figure 2.13 A parallel structure.

2.11.2 Parallel Structures

In a parallel structure, a system is composed of N independent sub-components, such that all of them must fail for the complete system to fail. A parallel interconnection can be schematically represented as in Figure 2.13.

The reliability of a parallel structure is better analyzed by considering the complementary unreliability function $Q(t)$. Given that all components must fail for the complete system to fail, we have that

$$Q(t) = \prod_{i=1}^{N} Q_i(t) = \prod_{i=1}^{N}(1 - R_i(t)) \tag{2.26}$$

By Equation (2.1) we have that

$$R(t) = 1 - \prod_{i=1}^{N}(1 - R_i(t)) \tag{2.27}$$

In symbols, this can be compactly written using the co-product operator as follows:

$$R(t) = \coprod_{i=1}^{N} R_i(t) \tag{2.28}$$

When all components have constant failure rates (exponential distribution with parameters λ_i), the reliability is given by

$$R(t) = 1 - \prod_{i=1}^{N}(1 - e^{-\lambda_i t}) \tag{2.29}$$

Notice that because we cannot write Equation (2.29) in the form of Equation (2.15), the failure rate of a parallel structure is not constant, even though the failure rates of individual components are.

References

Bozzano, M. and A. Villafiorita (2007). The FSAP/NuSMV-SA safety analysis platform. *Software Tools for Technology Transfer* 9(1), 5–24.

FSAP (Last retrieved on November 15, 2009). The FSAP/NuSMV-SA Platform. Available at `https://es.fbk.eu/tools/FSAP`.

Gonyeau, J. (2009). Chernobyl Event. Available at `http://www.nucleartourist.com/`. Last retrieved on November 15, 2009.

Griffin, S. (2009). Internet Pioneers. Available at `http://www.ibiblio.org/pioneers/index.html`. Last retrieved on November 15, 2009.

IBM (1964). Laboratory Maintanance Instructions—Saturn V Launch Vehicle Digital Computer. Available at `http://ntrs.nasa.gov/archive/nasa/casi.ntrs.nasa.gov/19730063841_1973063841.pdf`. Last retrieved on November 15, 2009.

Laprie, J. (1991). *Dependability: Basic Concepts and Terminology*. Berlin: Springer.

Lee, S., J.-il Jung, and I. Lee (2007). Voting structures for cascaded triple modular redundant modules. *IEICE Electronics Express* 4(21), 657.

Lee, S. and I. Lee (2001). Staggered voting for TMR shift register chains in poly-Si TFT-LCDs. *Journal of Information Display* 2(2), 22–26.

Leveson, N.G. (1995). *Safeware: System Safety and Computers*. Reading, MA: Addison-Wesley.

Miller, S.P. and M.W. Whalen (2005). A Methodology for the Design and Verification of Globally Asynchronous/Locally Synchronous Architectures. Technical Report CR-2005-213912, NASA.

Moore, E.F. and C.E. Shannon (1956). Reliable circuits using less reliable relays. *Journal of the Franklin Institute* 262, 191–208 and 281–297.

Rausand, M. and A. Høyland (2004). *System Reliability Theory: Models, Statistical Methods, and Applications*. New York: Wiley.

Siewiorek, D.P. and P. Narasimhan (2005). Fault tolerant architecture for space and avionics applications. In *Proc. 1st International Forum on Integrated System Health Engineering and Management in Aerospace (ISHEM 2005)*.

Storey, N. (1996). *Safety Critical Computer Systems.* Reading, MA: Addison-Wesley.

Various Authors (1987). Report on the Accident at the Chernobyl Nuclear Power Station. Technical Report NUREG-1250, Nuclear Regulatory Commission.

Various Authors (2009a). Chernobyl disaster. Available at `http://en.wikipedia.org/w/index.php?title=Chernobyl_disaster`. Last retrieved on November 15, 2009.

Various Authors (2009b). The history of the internet. Available at `http://en.wikipedia.org/wiki/History_of_the_Internet`. Last retrieved on November 15, 2009.

Von Neumann, J. (1956). Probabilistic logics and the synthesis of reliable organisms from unreliable components. In C.E. Shannon and J. McCarthy (Eds.), *Automata Studies*, Number 34, pp. 43–98. Princeton, NJ: Princeton University Press.

Chapter 3

Techniques for Safety Assessment

3.1 Introduction

There are several techniques to evaluate and improve the safety of a complex system. However, preliminary to any kind of evaluation or improvement is the identification of the *hazards*, that is, the events that are safety relevant. A **hazard** can be defined as any condition that has potentially harmful consequences for people or the environment. For instance, a train going through a level crossing when the gates are open and a leakage of radioactive material from a nuclear plant are both examples of hazards. A hazard that results in harmful consequences can be termed an **accident**. An accident is initiated by an **accidental event**, for instance, leakage in the example of the nuclear power plant. Not all safety hazards will necessarily result in accidents, however. For example, a train crossing when the gates are open will not necessarily cause an accident. The occurrence of such an event is called an **incident** (alternatively, **near accident** or **near miss**), to denote its potential to become an accident. The identification, classification, and reduction of the hazards are the objectives of hazard analysis, which is discussed in Section 3.2.

The evaluation of the possible safety hazards does not exhaust the preliminary analyses that are needed to deal with safety management in critical systems. As it is not possible to achieve absolute safety, designers of a critical system will necessarily be confronted with making decisions about which level of safety may or may not be considered acceptable for the system at hand. To make these decisions, two factors must be evaluated. First, a hazard must be evaluated in terms of the **severity** of the possible consequences that it may have. For instance, a hazard that has the potential of causing an accident where human lives may be lost can be deemed more serious than one from which only minor injuries may result. Second, a hazard must be evaluated in terms of the likelihood, or probability, of its occurrence. For instance, the probability of an airplane being struck by a meteorite will be lower than that of colliding with another airplane. By combining the notion of severity and probability of occurrence of a hazard, we can define the notion of **risk**. A risk can be thought of as the combination of the severity related to the potential consequences of a safety hazard, and the probability of its occurrence. Risk analysis is discussed in Section 3.3.

3.2 Hazard Analysis

The objective of hazard analysis is the identification and assessment of all the hazards that can result from the operation of a given system, and the selection of the means of controlling or eliminating them. Hazard analysis is an integral part of the design of safety-critical systems, and it assists the decision-making process throughout the design life cycle, from the early design stages to the construction, testing, operation, and maintenance of a system.

Traditionally, several techniques are used to perform hazard analysis. In fact, hazard analysis can be seen as a collection of different techniques that can be used independently, or under different circumstances, and that can give complementary insights into the potential safety problems of a given system.

A distinction can be made between **inductive** and **deductive** techniques. Inductive approaches typically start by considering the initiating causes of a given hazard, and trace them forward, through event propagation, to the corresponding safety consequences. Deductive techniques, on the other hand, start by considering an unintended behavior of the system at hand, and trace it, in a backward reasoning fashion, to the corresponding causes. An example of deductive technique is Fault Tree Analysis (FTA), whereas Failure Mode and Effects Analysis (FMEA) is an example of inductive approach. In the remainder of this section we discuss some of the most well-known techniques for hazard analysis.

3.2.1 Fault Tree Analysis (FTA)

Fault Tree Analysis (FTA) was first developed by Bell Telephone Laboratories during the early 1960s for the U.S. Air Force's Minuteman Launch Control System (Intercontinental Ballistic Missiles and Bombers), and later used extensively for U.S. nuclear power plants and by the Boeing Company. Today it is commonly used in all major fields of safety engineering. FTA can be described as a deductive, analytical technique, whereby an undesired state (the so-called **top-level event** (TLE)) is specified, and the system is analyzed for the possible chains of **basic events** (typically, system faults) that may cause the top event to occur. A Fault Tree (FT) is a systematic representation of such chains of events, which makes use of logical gates, corresponding to logical connectives such as AND and OR, to depict the logical

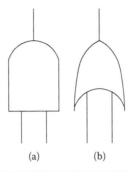

(a)   (b)

Figure 3.1   Basic fault tree gates: (a) AND gate, and (b) OR gate.

interrelationships linking the basic events with the top event. An AND gate relates events that are both required to occur to cause the hazard, whereas an OR gate represents alternative causes. The basic events are the leaves of the FT, whereas events that appear in between the root and the leaves are called **intermediate events**. The tree is typically drawn with the TLE at the top of the diagram.

The set of basic gate and event symbols is depicted, respectively, in Figure 3.1 and Figure 3.2. Here, the undeveloped event symbol is used to mark events that are not further developed because they are not relevant for the analysis or because of insufficient information. A rich source of information about FTs and their construction is the *Fault Tree Handbook* (Vesely et al., 1981) and its follow-up Vesely et al. (2002). According to Vesely et al. (2002), an event (either the TLE or an intermediate one) must be developed considering the *immediate, necessary, and sufficient causes* for its occurrence. Causes that are considered elementary faults are developed as basic events, whereas the remaining causes are developed as intermediate events. This rule applies recursively to the newly generated intermediate events, which must in turn be traced back to their causes, until the tree is completely developed. In Vesely et al. (2002), a plethora of symbols, not illustrated in this book, are presented. For instance, **transfer symbols** may be used to link different (parts of) fault

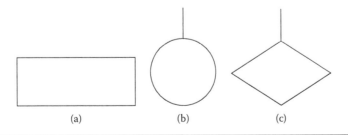

(a)   (b)   (c)

Figure 3.2   Basic fault tree events: (a) intermediate event, (b) basic event, and (c) undeveloped event.

trees. Moreover, **inhibit gates** and **conditioning events** can be used to constrain the ways that faults are propagated inside the FT. Finally, dynamic gates such as the **priority AND** gate may be used to enforce temporal constraints on the occurrence of the input events. Important notions related to the development of an FT are the **scope** and **boundary** of the analysis, and the **level of resolution**. The scope and boundary define which parts of the system will be included in the analysis, which events are to be considered basic events, and under which hypotheses and operational conditions the system will be analyzed. Moreover, the boundary conditions define the initial state of the system and the assumptions on the surrounding environment. For instance, depending on the chosen scope and boundary, the analysis may be performed at the system or sub-system level. If an event is outside the scope of the analysis, it is considered a basic event and not further developed. The resolution of the analysis defines the level of detail used to trace back an event to its causes. The level of resolution dictates which events must be traced back to more primitives causes and which ones can be left undeveloped. For instance, a valve malfunction can be considered a basic event, or traced back to the failure of one or more of its mechanical sub-components. Very often, when a quantitative evaluation of an FT is desired, the level of resolution of the analysis is dictated by the ability to precisely foresee the probability of occurrence of the basic events. In some cases, abstraction and refinement techniques can be used to progressively increase the level of resolution. Ultimately, it is the responsibility of the safety engineer to decide the boundary and level of resolution of the analysis.

In general, there is no unique way an FT can be built. In particular, there may be different choices for the intermediate events, and different ways to develop them. The guidelines given in Vesely et al. (2002) distinguish the case where a fault is localized to a given component ("state of component" fault) from the case where it is not ("state of system" fault). In the latter case, a fault is developed by considering its immediate, necessary, and sufficient causes. This may involve investigating the faults of multiple components (a notable example is for a fault-tolerant system containing redundant components). If the fault is localized to a given component, then its *primary*, *secondary*, and *command* faults are investigated. Primary and secondary faults differ, depending on whether the fault occurs in an environment for which the component is qualified (primary fault) or not qualified (secondary fault), whereas a command fault is due to a proper operation of a component, but at the wrong time or in the wrong place. The guidelines in Vesely et al. (2002) also suggest that the derivation of the immediate causes works by considering the **failure modes** and **failure mechanisms** (i.e., the way a failure mode can occur) of components. The immediate causes of a system failure are failure mechanisms for the system, that, in turn, are failure modes from the perspective of its sub-components. Hence, in this view, the FT is developed by following the system hierarchy and turning attention from mechanisms to modes during the development, until the desired level of resolution of the fault tree is reached.

A simple example of fault tree is shown in Figure 3.3.

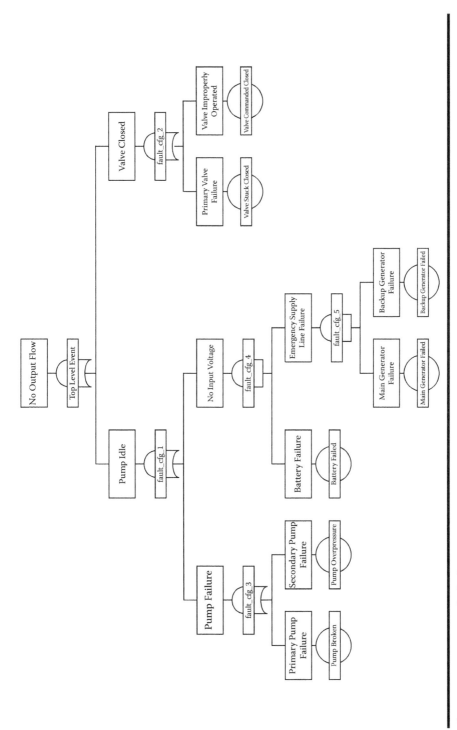

Figure 3.3   An example fault tree.

It is important to remark that what makes the FT useful is not just the logical relationship between the basic events and the top event (in terms of the underlying Boolean function), but rather the way these events are connected (which is related to the notion of *causality*), together with a proper choice of intermediate events that are semantically relevant for safety engineers. As an exemplification, consider the FT depicted in Figure 3.4. The fault tree contains three basic events, A, B, and C, and five intermediate events, $E1$ through $E5$.

In logical terms, this fault tree can be represented by the following logical formula:

$$A \wedge (A \vee (B \vee C)) \wedge (C \vee (A \wedge B)) \tag{3.1}$$

with the intended meaning that a propositional symbol is true whenever the corresponding basic event occurs. Hence the top event occurs if and only if the formula evaluates to TRUE. In strictly logical terms, fault trees can be considered equivalent if the associated logical formulae are equivalent. For instance, it is straightforward to see that the above formula is logically equivalent to

$$(A \wedge B) \vee (A \wedge C) \tag{3.2}$$

which can be graphically represented by the FT in Figure 3.5. This shape is of particular interest in reliability analysis, in that it represents the occurrence of a top event in terms of the so-called *Minimal Cut Set* (MCS). An MCS can be seen as the smallest combination of component failures that can cause the top event to occur. Both the previous trees have the same MCS, that is, $(A \wedge B)$ and $(A \wedge C)$.

Logically, a minimized FT such as the one in Figure 3.5 is associated with a Boolean formula in disjunctive normal form (i.e., a disjunction of conjunctions of propositional symbols). MCSs are of particular interest in reliability analysis because they represent simpler explanations for the top event. MCSs containing only one basic event are single points of failure that alone can cause the top event; hence they typically represent a weak point in the design of the corresponding system. MCSs are typically used as a starting point for quantitative analysis.

We conclude this section by discussing qualitative and quantitative evaluation of FTs. The fault tree model is not in itself a quantitative model, but rather a qualitative model that can be evaluated quantitatively, specifically to determine the probability of a safety hazard. Quantitative evaluation of an FT requires, first of all, an estimation of the probability of occurrence of the basic events. This estimation can be done in several ways. For instance, the likelihood of a fault of a piece of hardware can be estimated on the basis of statistical data about faults of the same sort of component. Statistical analysis can be used to predict software faults as well, although the inherent characteristic of software faults being systematic faults, makes this sort of estimation more difficult. Evaluation of the FT is performed from the bottom up to the top of the tree. Once an estimation of the fault probability for basic events is given, evaluating the probability of occurrence of the intermediate events of the tree is

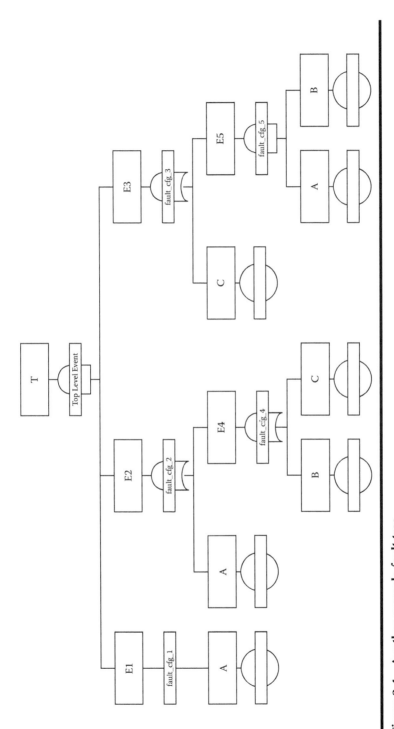

Figure 3.4 Another example fault tree.

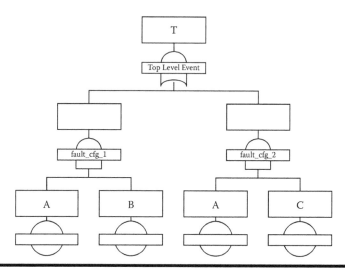

Figure 3.5   The fault tree in Figure 3.4 minimized.

carried out by combining the probabilities of the underlying parts of the tree that have already been evaluated. Combination of probabilities may involve evaluating conditional probabilities, when there are dependencies between different events.

Tools for FT evaluation include Galileo (Sullivan et al., 1999), which is based on the DIFTree (Dynamic Innovative Fault Tree) methodology for the analysis of dynamic fault trees (Dugan et al., 1992; Manian et al., 1998). This methodology is able to identify independent sub-trees, translate them into suitable models, analyze them, and integrate the results of the evaluation. The Galileo tool allows the user to edit a fault tree in a textual or graphical format, and provides several alternative techniques for fault tree evaluation.

3.2.2 Failure Mode and Effects Analysis (FMEA)

Failure Mode and Effects Analysis (FMEA) is a classical, inductive technique to perform hazard analysis. It was first introduced in the late 1940s in a military context by the U.S. Armed Forces, and later used in aerospace applications such as the Apollo program by NASA. The use of FMEA has spread extensively since then, and is nowadays common in a variety of domains.

FMEA starts with the identification of the failure modes of the components of the system under investigation and, using forward reasoning, assesses their effects on the complete system. As for FTA, the analysis can be performed at different levels. Typically, the failure modes of the components at a given level are considered, and the objective of the analysis is to identify the effects of the failure modes on that level and, usually, on the next higher level of the design. The concepts of scope and boundary of the analysis described in Section 3.2.1 apply to FMEA as well, and are typically decided by the safety engineer, taking into account user requirements. Finally, FMEA can be applied at the hardware component level, or at the functional level, that is, considering the functional behavior of each component instead of its hardware implementation.

Typically, FMEA considers only single faults, although combinations of faults can be considered in particular cases. For instance, the European Space Agency (2001) prescribes that when a single item failure is nondetectable (e.g., due to the presence of a redundancy device), the analysis will be extended to assess the effects of additional failures, occurring in the relevant redundancy item, which in combination with the first failure can lead to catastrophic or critical consequences at system level. Similarly, the failures of warning or emergency devices should be analyzed together with the failures whose occurrence is monitored by these devices.

An extension of FMEA is Failure Modes, Effects and Criticality Analysis (FMECA). With respect to FMEA, FMECA also takes into account the criticality of the consequences of component failures, which is computed on the basis of their severity and their probability or frequency of occurrence (compare Section 3.3). FMECA can also be applied to identify weaknesses in the development processes (e.g., in the assembly or manufacturing) of a given product. Such analysis goes under the name of *process FMECA* (European Space Agency, 2001).

The results of FMEA are recorded in a so-called FMEA table. FMEA tables may assume several different forms. An FMEA table is structured in different entries, each entry recording the information related to the effect of a given failure on the system. Typically, information that is always present includes the definition of the failure mode and the corresponding effects on the systems (which can be distinguished into local effects and system effects). Moreover, additional information can include a severity classification of each identified failure mode, the operational mode (or phase) in which the failure can occur, a list of possible causes of the failure, the failure detection means, and possible corrective actions to be taken to prevent or limit the effects of the failure. As for FTA, FMEA can be performed qualitatively or quantitatively. In the latter case, failure rates are attached to each failure mode and used to compute the probability of the failure effects. A simple example of FMEA table is presented in Table 3.1.

The results of FMEA can be summarized in a table called Failure Modes and Effects Summary (FMES) (SAE, 1996), in which failure modes leading to the same failure effects are grouped together. An FMES can be used as an input to FTA or other analyses.

3.2.3 Hazard and Operability (HAZOP) Studies

The HAZOP technique is another example of inductive method to perform hazard analysis. It was first developed within the chemical domain by ICI in the 1960s, and is nowadays most notably used in the process industries, such as the chemical, petrochemical, and nuclear industries, although it may also be used in other domains.

Table 3.1 An Example FMEA Table

Ref. No.	Item	Failure Mode	Failure Cause	Local Effects	System Effects	Detection Means	Severity	Corrective Actions
1	Pump	Fails to operate	Comp. broken	Coolant temperature increases	Reactor temperature increases	Temperature alarm	Major	Start secondary pump
			No input flow					Switch to secondary circuit
2	Valve	Stuck closed	Comp. broken	Excess liquid	Reactor pressure increases	Coolant level sensor	Critical	Open release valve
3		Stuck open	Comp. broken	Insufficient liquid	Reactor temperature increases	Coolant level sensor	Critical	Open tank valve

HAZOP is based on a team approach to hazard analysis, with the idea that a team of experts will be able to identify more problems than a set of individuals working separately. It is typically carried out by a team comprising several engineers with different backgrounds and competencies (for instance, experts of the domain and engineers with extensive training in hazard analysis techniques). HAZOP is typically conducted in a number of different sessions, during which potential problems are identified and the process under analysis is thoroughly reviewed by the team of experts.

The objective of HAZOP is to investigate the basic set of operations of the system under analysis, consider the possible deviations from normal operation, and identify their potential hazardous effects. As for FMEA, once hazards have been identified, it is possible to suggest corrective actions on the system that might help in preventing them or reducing their impact. HAZOP starts from the identification of parts of the process called *nodes*. Each node typically has an associated process *parameter* and a design *intention* that states the operational conditions under which the process must take place for correct operation. For instance, if we consider the flow of a chemical substance through a pipe, from a source tank to a destination, the parameter could be the flow itself or a characteristic of the flow, for example, the temperature or pressure of the substance while flowing through the pipe. The design intention would involve, for example, the amount of substance to be transported through the pipe, or pose constraints on the temperature and pressure of the substance.

Once the parameters and corresponding design intentions have been identified, the HAZOP team focuses on the identification of all the possible *deviations* from the design intentions, and in the investigation of their impact on the final system. For each node and parameter, the analysis is repeated for each possible deviation that has been identified. The possible deviations from normal operation are typically classified formally on the basis of a number of *guide words* that qualify the parameter. For instance, in the chemical industry guide words such as *no* (negation of the intention), *more* or *less* (an increase or decrease in the amount of a physical entity), and *reverse* (the opposite of the design intent) can be used as prefixes of the corresponding parameters (e.g., the guide word *no* combined with the parameter *flow* results in the deviation *no flow*). In the previous example of the flow, these guide words can be used to denote, respectively, the absence of flow, an increase or decrease in the amount of substance that is flowing, and reverse flow. In other domains, such as computer-based systems or software systems, guide words such as *early, late, before,* or *after* can be used to state temporal deviations with respect to the design intention. The guide words and their meaning are generally defined depending on the specific domain in which HAZOP is carried out. The results of HAZOP are typically recorded in a table, where each entry contains a specification of the parameters and deviations analyzed, together with a description of the relevant causes, the consequences on the system, and possibly corrective actions suggested by the HAZOP team in order to reduce risk to an acceptable level. A simple example of HAZOP table is presented in Table 3.2.

Table 3.2   An example HAZOP Table

Ref. No.	Item	Parameter	Guide Word	Causes	Consequences	Corrective Actions
1	Pump	Flow	No	Pump failure	Insufficient tank level	Close output valve
2			Less	Sensor failure	Insufficient tank level	Close output valve, replace sensor
3			More	Sensor failure	Excess liquid in tank	Open release valve, replace sensor
4	Sensor	Voltage	No	Supply line failure	Incorrect reading	Switch to secondary line
5			Less	Sensor failure	Incorrect reading	Replace sensor
6			More	Current regulator fault	Possible damage to sensor	Replace current regulator

3.2.4 Event Tree Analysis

Another example of inductive technique for hazard analysis, which proceeds in a bottom-up fashion, is Event Tree Analysis (ETA). ETA was first used in the 1960s within the nuclear industry but is now also utilized in other domains, such as in different process industries and in transportation.

ETA can be considered an alternative with respect to other classical techniques such as FTA and FMEA. ETA starts from an *initiating event*, typically drawn at the left of the diagram, and proceeds from left to right, branching on further events that are identified during the analysis, to determine the possible consequences on the system. Typically, binary branching is used; that is, either an event or its complement is assumed to occur (e.g., success or failure of a sub-system that is supposed to intervene in response to the event identified in the previous layer of the tree). The tree is developed until the desired consequences on the system, called *end events* or *outcome scenarios*, have been reached. As a difference compared with FTA, the events may correspond either to expected operations of the system under analysis (e.g., opening or closing of a valve), or to fault conditions (e.g., a valve failing stuck closed). Given that each event produces a branch in the diagram, the complete tree

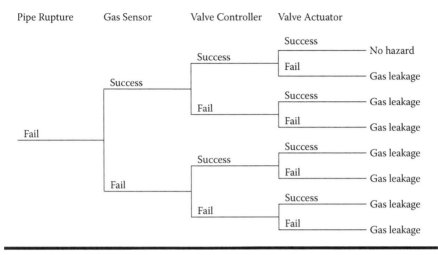

Pipe Rupture   Gas Sensor   Valve Controller   Valve Actuator

Figure 3.6   An example event tree.

will contain 2^N branches, with N being the number of events. For this reason, and given that both normal and hazardous operations are considered in the analysis, event trees can result in very large trees for complex systems. A simple example of event tree is shown in Figure 3.6.

Similar to fault trees, event trees can be quantified. This is done by assigning probabilities to each branch of the tree, and combining the probabilities of each branch to obtain a probability for each outcome scenario. Under the hypothesis of independent failures, the probability of an outcome scenario is computed by multiplying the frequency of occurrence of the initiating event by the probabilities of the branch points encountered along the corresponding branch.

More recently, event trees have been extended to take into consideration dynamic aspects such as the timing and order of intervention and failure of the systems under analysis. The resulting methodology goes under the name of Dynamic Event Trees (DETs) (Cojazzi et al., 1992).

3.3 Risk Analysis

The notion of risk combines a measure of the severity of the consequences of a safety hazard and a measure of the likelihood (probability or frequency) of its occurrence. A further characterization of the notion of risk is that it always refers to undesired future consequences, it involves an expectation of loss of some form (loss of human lives, economic loss), and it presents an element of uncertainty. Different hazards can be associated with different degrees of risk. For instance, a hazard with potentially catastrophic consequences could be deemed acceptable if its occurrence is extremely improbable. Conversely, a hazard with minor consequences could be deemed unacceptable if it is expected to happen very frequently. In this chapter we discuss how hazards and risks can be measured and classified, what makes a risk acceptable, and how risks can be managed.

3.4 Risk Measures

Risks can be measured in different ways, either qualitatively or quantitatively. When measured quantitatively, risk is defined on the basis of probability measures and of quantitative measures of the potential consequences of hazards. The latter measures strongly depend on the kind of consequences that are considered when evaluating the specific hazards. For example, measures can be defined in terms of the number of fatalities (when considering loss of human lives) or the additional money that is spent as a consequence of an accident resulting from a hazard (when considering economic loss). In some contexts, the evaluation of the harmful consequences may be more difficult (e.g., evaluation of environmental damage, damage to reputation).

An example of quantitative risk measure (Rausand and Høyland, 2004), tailored to consequences that involve loss of lives, is the **individual risk per annum** (IRPA), which measures the probability that an individual dies within a 1-year exposure to a given hazard. This risk can be also expressed, as a safety performance measure, as the ratio between the number of observed fatalities and the total number of employee/years exposed. The individual risk per annum varies considerably in different industrial sectors, depending on the hazardous activity under consideration. Moreover, for some kinds of hazards (e.g., exposure to harmful emissions), it is possible to quantify the risk, depending on the physical distance from the source of the hazard (the plant). Sometimes, other safety performance measures, such as the number of **deaths per million**, can be used. Another measure is called the **fatal accident rate**; it measures the expected number of fatalities per 10^8 hours of exposure. When risk is measured qualitatively, severity and frequency of occurrence are classified into a number of different classes. This is the approach that is typically followed within most standards. The classifications can vary considerably, depending on the industrial sector and the specific standard. We discuss the classification of hazards in terms of severity and frequency in more detail in the next two sections.

3.4.1 Classification of Hazards: Severity

All standards typically classify hazards into a number of classes that take into account the severity of the consequences of potential hazards. The classification uses names such as *catastrophic, critical, hazardous,* and *negligible* to denote classes with different degrees of severity, from the most severe to the least severe. The precise classification varies, depending on the standard.

As an example, civil aviation authorities in Europe and the United States organize hazards into five different classes, named *Catastrophic, Hazardous* (*Severe-Major*), *Major, Minor,* and *No Safety Effect.* A hazard is associated with a specific class, depending on the (qualitative) observable consequences associated with the hazard. For example, the highest category, *Catastrophic,* is associated with failure conditions that would prevent safe flight and landing. The other classes are associated with a different degree of reduction in the safety margins or functional capabilities of the aircraft.

Other standards classify hazards depending on the potential accident severity. For instance, the Defence Standard 00-56 (U.K. Ministry of Defence, 2007) of the U.K. Ministry of Defence organizes hazards into four different classes, called *Catastrophic, Critical, Marginal,* and *Negligible.* Each class is associated with the number of potential fatalities or injuries that can result as a consequence of hazard exposure, with the highest class being associated with multiple deaths and the lowest one with a single minor injury.

3.4.2 Classification of Hazards: Frequency

Similar to the severity classification, hazards are also classified depending on the frequency of their occurrence. As for severity, specific standards provide different classifications. The classification uses names such as *frequent, probable,* and *remote* to denote different degrees of frequency. Often, each class is also characterized quantitatively by associating it with a range of probabilities of occurrence. The probability can be expressed in different units, depending on the standard. For instance, in avionics it is typical to express it as number of events per flight-hour. In other domains, events per year of operation might be used. For protection systems (e.g., shutdown system for nuclear power plants), it is common to express the probabilities in terms of *failure on demand,* that is, the ratio between the number of failures and the number of times the system is operated.

As an example, civil aviation authorities in Europe and the United States use the following classification: *Probable* (further divided into *Frequent* and *Reasonably Probable*), *Improbable* (further divided into *Remote* and *Extremely Remote*), and *Extremely Improbable.* Each class is associated with a different probability (number of events per flight-hour) ranging from 1 to 1e-9. Sometimes, as in Defence Standard 00-56 (U.K. Ministry of Defence, 2007), the probability classes are associated with qualitative descriptions such as *likely to occur often* and *likely to occur several times.*

3.4.3 Classification of Risks

When severity and frequency of hazards are classified qualitatively, as seen in Section 3.4.1 and Section 3.4.2, their measures can be combined to produce a classification of the risk associated with the hazards. This classification is called **risk class**, or **risk level**. Risk classes can also be used when severity and frequency are evaluated quantitatively, in order to simplify the presentation of the risks. Similar to severity and frequency, specific standards provide their own risk classification.

Risk classes can be associated with different design and development guidelines. As an example, avionic standard SAE ARP4761 (SAE, 1996) associates each risk class with a different **Development Assurance Level** (DAL). There are five levels, named using letters from A to E.

See Section 4.9.2 for a discussion of how the classification of hazards according to severity and probability determines the required assurance level for the development of a system component.

3.4.4 Risk Management and Acceptance

The goal of risk management is to reduce the likelihood of potential accidents and mitigate their consequences. This goal can be accomplished in different ways, for instance by eliminating potential hazards, by preventing the occurrence of accidental events, or by reducing the effects of accidents. Risk management therefore involves several activities such as hazard identification and assessment, risk evaluation, and risk reduction. Moreover, an essential part required for the development of safety-critical systems is the production of a safety argument that supports the risk management activities and demonstrates that the risks have been reduced to an acceptable level. Such an argument is typically required by certification authorities.

There is no practical definition of what constitutes an acceptable risk. Ultimately, the decision about which level of risk may be acceptable relies on several criteria and depends on the industrial sector and the specific application at hand. Within the same application, acceptability of risk may depend on the specific hazard being considered. Typically, one carries out a cost-benefit analysis of the level of risk and the effort needed to further reduce the existing risks. Sometimes, the decision about acceptability of risk relies on professional and expert judgment. When certification is required, the final decision relies on the certification authority.

In Defence Standard 00-56 (U.K. Ministry of Defence, 2007), risks are classified into three levels: *Unacceptable, Tolerable*, and *Broadly Acceptable*. The latter category comprises risks that can generally be tolerated, provided the risks have been reduced *wherever reasonably practicable*. The category of tolerable risks comprises risks that can be tolerated if they have been demonstrated to be **ALARP (As Low As Reasonably Practicable)**. A risk is said to be ALARP if it has been demonstrated that the cost of further risk reduction (in terms of money or effort) is *grossly disproportionate* with respect to the reduction gained. Hence, demonstrating that a risk is ALARP involves a cost-benefit evaluation. In general, proving that risks have been mitigated is not enough.

3.4.5 Safety Integrity Levels

We conclude this section on risk analysis by discussing the notion of **safety integrity levels**. In the context of risk analysis, safety integrity refers to the likelihood that a system will perform all its safety functions in a satisfactory way, with respect to given operational conditions and period of time (compare Section 2.2.4). Sometimes, safety integrity is further classified into **hardware integrity**, **systematic integrity**, and **software integrity**, which refer, respectively, to hardware faults, systematic faults, and software faults.

As for hazard and risk classification, safety integrity levels are very often characterized qualitatively. Different standards rely on different classifications. Each safety integrity level is often associated with quantitative requirements about the failure rates (e.g., failure per year, or failure on demand for protection systems) that are acceptable within that class.

The notion of safety integrity level, for a specific application, is related but orthogonal with respect to the risk classification discussed in Section 3.4.3. Risk classes categorize the likelihood and consequences of hazardous events related to the application, whereas safety integrity is a measure of the likelihood that the application will perform its tasks safely. The allocation of an application to a specific safety integrity level depends on a number of factors. It depends on the risk class of the application, but it also clearly depends on the level of criticality of the application. Criteria for assigning safety integrity levels are usually described within industry-specific standards. Typically, the standards also define the design and development methodologies that should be followed to ensure that the system being developed meets the requirements of a specific safety level. Finally, as for acceptability of risks, development of a safety-critical application also involves formal arguments to demonstrate that a certain integrity level has been reached.

References

Cojazzi, G., J.M. Izquierdo, E. Meléndez, and M.S. Perea (1992). The reliability and safety assessment of protection systems by the use of dynamic event trees. The DYLAM-TRETA package. In *Proc. XVIII Annual Meeting Spanish Nuclear Society*.

Dugan, J., S. Bavuso, and M. Boyd (1992). Dynamic fault-tree models for fault-tolerant computer systems. *IEEE Transactions on Reliability* 41(3), 363–377.

European Space Agency (2001). Space Products Assurance: Failure Modes, Effects and Criticality Analysis (FMECA). Technical Report ECSS-Q-30-02A, European Cooperation for Space Standardization (ECSS).

Manian, R., J. Dugan, D. Coppit, and K. Sullivan (1998). Combining various solution techniques for dynamic fault tree analysis of computer systems. In *Proc. 3rd IEEE International Symposium on High-Assurance Systems Engineering (HASE '98)*, pp. 21–28. Washington, D.C.: IEEE Computer Society.

Rausand, M. and A. Høyland (2004). *System Reliability Theory: Models, Statistical Methods, and Applications*. New York: Wiley.

SAE (1996). Guidelines and Methods for Conducting the Safety Assessment Process on Civil Airborne Systems and Equipment. Technical Report ARP4761, Society of Automotive Engineers.

Sullivan, K.J., J.B. Dugan, and D. Coppit (1999). The Galileo fault tree analysis tool. In *Proc. Symposium on Fault-Tolerant Computing (FTCS 1999)*, pp. 232–235. Washington, D.C.: IEEE Computer Society.

U.K. Ministry of Defence (2007). Defence Standard 00-56. Safety Management Requirements for Defence Systems. Part I: Requirements. II. Guidance on Establishing a Means of Complying with Part 1. U.K. Ministry of Defence.

Vesely, W., M. Stamatelatos, J. Dugan, J. Fragola, J. Minarick III, and J. Railsback (2002). Fault Tree Handbook with Aerospace Applications. Technical report, NASA.

Vesely, W.E., F.F. Goldberg, N.H. Roberts, and D.F. Haasl (1981). Fault Tree Handbook. Technical Report NUREG-0492, Systems and Reliability Research Office of Nuclear Regulatory Research U.S. Nuclear Regulatory Commission.

Chapter 4

Development of Safety-Critical Applications

4.1 Introduction

This chapter focuses on the development process of complex systems. Thus, we digress a bit from the main subject of this book (namely, the usage of formal techniques for safety assessment) and focus instead on aspects more closely related to the management and technical organizations of system development activities. There are, in fact, two very good reasons to move away from the techniques and take a look at a broader picture, in which fault trees, formal methods, and FMECAs are just one of the elements contributing to building safe systems.

The first reason is the importance the process has in building products that meet safety goals and stakeholders' expectations, in terms, for example, of functionality, performance, and quality. We have seen in Section 2.5 that there are four different approaches to managing faults: avoidance, removal, detection, and tolerance. Of these, only the last (fault tolerance) primarily (not exclusively!) relies on the adoption of a technical solution. The effectiveness of the others depends, in one way or another, on the development process.

The second reason is practical. Safety assessment and formal techniques come at a cost. Moreover, they are not equally effective during all stages of development. The safety effort and the application of formal techniques, therefore, must be properly tailored to achieve a satisfactory level of safety, while at the same time meeting practical considerations about time and costs. Understanding both formal methods and the development process is the only way to achieve this goal.

More generally, a sound development process has a positive impact not only on the resulting product, but also on the project and on the performing organization. From the project point of view, in fact, a sound process significantly increases the probability of delivering on time and within budget. From the organization point of view, process standardization is a key enabler for improving the efficiency and the efficacy with which the organization delivers. Finally, on top of the advantages mentioned above, a sound development process is an essential precondition for product certification.

This chapter looks at the way in which the criticality and complexity of a system determine the activities that must be adopted for its development. In particular, we will see how complexity entails a "recursive" organization of activities, while increasing levels of criticality call for higher levels of formality in the development. Such formality is achieved through the introduction of specific activities, for example, aimed at improving control over the development process and traceability. Certification of highly critical components in the aerospace sector, for instance, requires us to keep under configuration control several artifacts produced during development.

Safety-critical systems are the most diverse in terms of complexity, criticality, and market constraints. Different standards and processes have thus been defined and adopted in different domains. For instance, it is quite intuitive that the development and certification of an Anti-lock Braking System (ABS) of a car is subject to development processes and standards that are quite different from those adopted for building the controller of a nuclear plant, although human lives depend on both applications. In this chapter we do not stick to a specific standard or application area. Rather, we present a framework that is general enough to accommodate different classes of systems, while being detailed enough to provide concrete guidance.

The chapter is organized as follows. We start by characterizing the main sources of complexity in a system's development and the main factors influencing, positively or negatively, development (Sections 4.2 to 4.5). This allows us to introduce a general development framework that includes the most relevant disciplines and activities for the development of a safety-critical application (Section 4.6 to 4.11). The overview is supported by a brief discussion on tool support (Section 4.12). We conclude the chapter with a brief description of process improvement frameworks (Section 4.13).

4.2 What Makes a System Complex

As hinted in Section 1.1, there is no precise or formal definition of system complexity. Following the SAE ARP4754 (SAE, 1996) standard, we defined "complex" a system "whose logic is difficult to comprehend," providing only a qualitative characterization.

Various theories have been proposed about the nature of complexity in systems. We mention Dvorak (2009), which recaps the definitions given by Dörner (1997) and Perrow (1984). For the former, complexity is due to the "existence of many interdependent variables." For the latter, complexity emerges from a combination of "tight coupling" and "complex interactions."

We focus on three main factors that determine the complexity of a system and consequently, the complexity of its development:

1. **Functions to be performed.** The system might be required to perform a wide range of functions or tasks regulated by complex physical or control laws. The system might be required to implement a complex control logic.
2. **Operating environment.** The system might be required to operate under harsh environmental conditions or under tight timing constraints, or be required to process inputs from several different sensors and send outputs to many actuators.
3. **Criticality.** The system might be required to be fail operational or, in any case, implement safety functions that could make its logic more difficult to comprehend.

We hint in Section 4.4 as to how these factors influence system development.

4.3 What Makes the Development Complex

The system characteristics mentioned above are what Dvorak (2009) calls the **essential complexity** of a system because they "specify the essence of what a system must do without biasing the solution." Essential complexity, however, is only half the story, as various other factors contribute to the complexity of the development and to the quality of the resulting product. These factors typically depend on the characteristics of the project and of the performing organization(s). We mention:

■ Novelty
■ Schedule constraints
■ Team
■ Geographical distribution of the team
■ Organization's maturity
■ Tool support

Let us describe them in more detail.

4.3.1 Novelty

Competitive development of successful products often relies on the adoption of new technologies. Novelty might derive from two factors: the technology might be used for the first time ever—at least in a specific application domain—or it might be new for the organization that has decided to adopt it. In both cases, however, novelty brings uncertainties and risks that can significantly increase the complexity of the development.

For instance, the introduction of composite materials for aircraft structures has allowed the industry to build airplanes that consume less fuel and are more comfortable for passengers. However, this required aircraft makers to modify airframe production, assembly, and inspection techniques.

4.3.2 Schedule Constraints

The critical path of a plan is the sequence of activities on which the delivery date of a project depends. *Any* delay in an activity in the critical path and the project is late. Plans have only one critical path. Projects, however, might have several *nearly critical* paths, namely sequences of activities for which some *small* delay might cause the delivery date to shift.

When working under tight timing or staffing constraints, the number of nearly critical paths in the plan is very high, and maintaining the planned delivery date might become a challenge. Risks in such situations include the possibility of overlooking important details (e.g., by spending too little time analyzing issues that could arise from integrating components); shortening important activities (software testing, for instance, is often shortened to accommodate delays during development); and compromising a proper information flow in the project (on which the quality of systems often depend).

One strategy commonly adopted to deal with this issue consists of adding some slack time between activities, so that delays in one activity do not propagate throughout the plan. Often, however, there are no opportunities to stretch the project schedule in such a way because success might depend on conclusion within a hard deadline. Think, for instance, of delivering a system in time to exploit a favorable launch window for a spacecraft.

4.3.3 Team

Project success or failure significantly depends on team capabilities and the project manager's success in building teams that share common vision and goals.

4.3.4 Geographical Distribution

Geographical distribution of the project team can be a blessing or a curse.

The opportunities are clear. Geographic distribution can mitigate problems related to staffing constraints, for example, by allowing the organization to hire people not available locally. It can help reduce development costs. It allows for hiring talented people otherwise unavailable. It can be a strategy to increase productivity, by having teams that can work "round-the-clock" (think, for instance, of software development teams located in different time zones).

The disadvantages and the risks, however, are also very evident. The information flow can be significantly affected, management complexity increases, and conflicts among teams might arise. See, for example, Sommerville (2007), Anschuetz (1998), Hinds and Mortensen (2005), and Miller (2008) for a discussion of the matter.

A paradigmatic example of the impact that geographical distribution can have on the development process is the story of the A380, the super jumbo jet built by AIRBUS (2009). AIRBUS offices are located in different countries in Europe, and it is common for the company to have projects with geographically distributed teams. (The situation applies equally well to other producers.)

In the case of the A380, the design of the aft fuselage and of the tail were assigned to different AIRBUS offices. However, different policies related to software updates in Spain, France, Germany, and the United Kingdom resulted in the design teams of the A380 using two different versions of the same CAD tool, Catia© (Dassault Systemes, 2009). During the integration of the designs, the different way in which some of the data about electrical wiring was managed by the two versions of the CAD resulted in the designs not matching.

The A380 has about 500 kilometers of wires. To solve the problem, a significant amount of work had to be re-done, resulting in delays of several months and costs estimated in the millions of euros (MRY/Reuters, 2006).

4.3.5 Organization's Maturity

Capability Maturity Model® Integration (CMMI) is a widely known framework to measure and improve an organization's maturity and capability to develop systems or deliver services (CMMI Product Team, 2009). The model, which we describe in more detail in Section 4.13, defines different levels of capability and maturity, ranging from *initial*, in organizations with no maturity and no structured capabilities, up to *optimized*, when the organization is able to analyze its own performances and improve its efficiency and its efficacy.

The keystone to moving up the ladder in the CMMI model is the adoption of standardized production practices—that is, a sound development process—that are consistently applied throughout the organization.

Such organizationwide standards, in fact, can greatly help tame the complexity of system development, because, for instance, they allow us to focus on the problem at hand (rather than, for example, spending resources on defining project standards), help improve the efficiency and efficacy of personnel (training and experience derives by having worked in previous projects of the same organization), and allow us to capitalize on the experience of previous projects (e.g., quantitative and qualitative data of previous projects can be used for the project at hand).

4.3.6 Tools

Software tools have become increasingly useful and effective in supporting the information-intensive activities that characterize the development of complex systems. Tools support project managers in analyzing trends and making predictions about, for example, time and costs; they support engineers in modeling, designing, and simulating systems; they support implementation with integrated environments and CADs; and finally, tools support effective communication, storage, and traceability of information.

A suggestive example can be taken from Formula 1 (F1). The development and prototyping of F1 cars is extremely expensive and time consuming. To cut both, racing teams heavily rely on CAD designs that are simulated in "digital" wind tunnels. This allows one to evaluate and change various aerodynamic and structural features directly on the blueprints, before building actual prototypes for final tests. Results from software simulators are very accurate and typically anticipate those of the experiments performed on prototypes. We mention the case of Noran Engineering, which reports that the Minardi F1 team has used their CAD software for testing directly on CADs—rather than on prototypes—various general requirements regarding the global shape of the chassis, compliance with safety regulation, and other performance requirements, as well as static analysis, buckling (linear/nonlinear, especially on crash cones), and surface contact (roll-bar crush) (NEi Software Inc., 2009).

4.4 Measuring the Impact of Complexity

COCOMO (Boehm, 1981; Boehm et al., 2000) and Function Point (FP) (Albrecht, 1979; Jones, 2007) are two well-known techniques for estimating the effort required to develop a software system. Simplifying a bit, they try to predict the software development effort by measuring some characteristics of the system to be developed and of the organization developing it. Such measures, which can be collected early (ideally, at the beginning of a project), are then fed to models that output estimations of the effort and time needed for development. The models have been derived by analyzing data collected from several projects. Thus, even if these techniques apply to software systems only, they allow us to hint at the relative importance that the factors introduced in the previous section have in system development.

In the remainder of the section we focus on COCOMO because it explicitly takes into account both the characteristics of the system to be developed and those of the performing organization. In the FP technique, by contrast, the characteristics of the performing organization are encapsulated in a "productivity factor" that transforms the output of the model—an abstract measurement of the size of the system—into effort. The productivity factor is company specific and must be determined with specific measurements collected by the organization applying the method.

COCOMO is a family of models developed between 1981 and the beginning of 2000 by Barry Boehm and his group. All the models are based on a similar characterization of the relationship between software size and development complexity. The models differ in the level of detail they provide, in their precision, and in the number of projects used to derive the formulae of the model. The first model, COCOMO'81, used data from 63 projects, whereas 161 projects were considered for the development of COCOMO II, the last model. The manuals describing the models (COCOMO II, 2000; University of Southern California, 2000, 1994) and the software tools for applying the model are available from the Center for Systems and Software Engineering of the University of Southern California (University of Southern California, 2009).

The general form of the COCOMO models is:

$$Effort = A \cdot (SIZE)^{(B+SF)} \cdot EAF$$

where

Effort is the effort needed to develop a system.

SIZE is the predicted size of the system to be developed, expressed in SLOC, source lines of code. Notice that, in general, this piece of information is certain only after the development is complete. Various techniques, however, can be used to predict its value at the beginning of a project. We mention analogy with other projects and the FP method, from which the estimated SLOCs (source lines of code) can be obtained from conversion tables mapping function points into lines of code.

A and *B* are companywide constants. COCOMO provides values for both constants (e.g., *A* is 3.6 and *B* is 1.2 in COCOMO'81), although in general they should be customized by each company willing to apply the method.

SF, the *Scale Factor*, is a project-dependent factor that models the economies and diseconomies of scale in software development. Notice that SF has an exponential impact on effort.

EAF, the *Effort Adjustment Factor*, is a project-dependent factor that adjusts the efforts according to the characteristics of the project at hand. Notice that EAF has a multiplicative impact on effort.

Both *SF* and *EAF* are defined using more elementary parameters, describing product and project characteristics. One such parameter, for instance, is *database size*, which measures the size of the database that will be used by the application to be developed. Each parameter is assigned a qualitative score, typically ranging from *very low* to *very high*, using guidelines provided by the COCOMO manuals. The qualitative scores are then transformed to numeric values, through conversion tables defined by the model. For instance, a *low* score to *database size* corresponds, in the COCOMO II model, to 0.9. By contrast, a score of *very high* corresponds to 1.28. The score *nominal* (middle of the scale) is always assigned the value 1. More specifically, in the most recent edition of the model, *SF* is the sum of the numeric value assigned to five different parameters. These parameters measure the capabilities of the performing organization. They are:

■ **Precedentness,** namely, the experience the organization has in developing the system.
■ **Flexibility,** namely, constraints related to project execution and, in particular, schedule constraints.

- **Team,** team maturity, experience, and cohesion.
- **Risk resolution,** namely, how risks are managed in the project.
- **Process maturity,** namely, organization maturity, according to the CMMI model.

Computation of the effort adjustment factors (*EAFs*) is slightly more complex. The value is obtained by multiplying the numeric values assigned to parameters describing product and project characteristics. The number of parameters to evaluate, however, changes according to when, in the software development life cycle, the method is applied. COCOMO, in fact, provides an **Early Design Model** that can be applied when the requirements are defined, and a **Post-Architecture Model** that can be applied when the architecture has been defined. All EAF parameters belong to one of the following four classes:

1. **Product,** which is meant to measure the complexity of the system to be developed. The parameter measures aspects such as required software reliability, database size, product complexity, and required usability.
2. **Platform,** which is meant to measure aspects related to the platform in which the system will run. It includes aspects such as execution time constraints, main storage constraints, and platform volatility.
3. **Personnel,** which is meant to measure the capabilities and experience of personnel used in the project. The parameter is computed by looking at aspects such as analysts' and programmers' general experience, specific experience with the execution platform, language and tool experience, and the expected staff turnover.
4. **Project,** which is meant to measure some project related characteristics. It includes the communication and team geographical distribution.

The sources of complexity presented in the previous section thus have their counterparts in the COCOMO method. Such correspondence is given in Table 4.1, in which we also highlight whether the parameter belongs to the *SF* or *EAF* class, that is, whether it has a linear or exponential effect on effort.

One simple way to understand the relative influence that each parameter has in determining the development effort is to look at the range of values it can assume. Minimum and maximum values of *SF* and *EAF* are shown, respectively, in Table 4.2 and 4.3. Figure 4.1 shows, in detail, the values of the RELY (reliability) parameter, because it is of particular interest in this book.

RELY is an *Effort Adjustment Factor* that measures the criticality of the software to be developed. As can be seen from Figure 4.1, according to the model, the effort required for critical applications (score of *very high* to RELY) is nearly twice that necessary for applications with *very low* criticality ($1.39/0.75 = 1.85$). This provides a rough estimation of the impact that criticality has in software development.

Table 4.1   Sources of Complexity and COCOMO Method

Sources of Complexity		COCOMO
Functions to be performed	EAF	Product (Parameter CPLX)
Operating environment	EAF	Platform
Criticality	EAF	Product (Parameter RELY)
Novelty	SF	Precedentness
Constraints on schedule	SF	Project, flexibility
Team	SF	Team
	EAF	Personnel
Geographical distribution	EAF	Project
Organization's maturity	SF	Process maturity
Tools	EAF	Project

Table 4.2   Weight of Different COCOMO Scale Factors (*SF*)

Factor	Best Case	Worst Case
Precedentness	0.01	0.06
Flexibility	0.01	0.05
Risk resolution	0.01	0.07
Team	0.01	0.05
Process maturity	0.01	0.07

Table 4.3   Weight of Different COCOMO Effort Multipliers Factors (*EAF*)

Factor	Best Case	Worst Case
Product	0.41	4.28
Platform	0.87	3.09
Personnel	0.25	4.28
Project	0.62	2.04

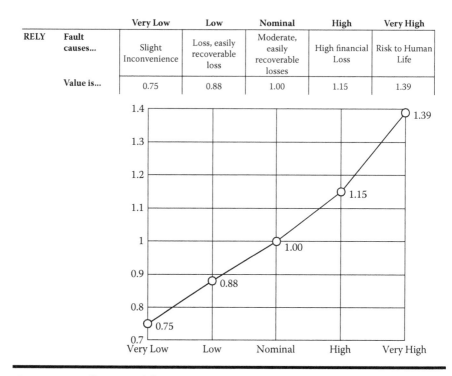

RELY	Fault causes...	Very Low	Low	Nominal	High	Very High
		Slight Inconvenience	Loss, easily recoverable loss	Moderate, easily recoverable losses	High financial Loss	Risk to Human Life
	Value is...	0.75	0.88	1.00	1.15	1.39

Figure 4.1 **Influence of the RELY parameter on software development.**

Table 4.2 and Table 4.3 can also be used as a decision taking tool because they allow us to determine the advantages or disadvantages that might be obtained by improving the score in one area. For instance, a score of *very low* for the *Personnel* parameter causes the effort to be four times (to be precise, 4.28 times) larger than that obtainable with a team score *nominal* (value: 1). Thus we can decide whether we can afford a team with less experience (which could presumably cost less) or whether a team with more experience is needed to achieve project goals.

Notice, finally, that we have different degrees of control over the factors described above. For instance, often we have little or no control over the *Product* factors because these are typically constrained by external conditions. We might have some control over the *Platform* factors (i.e., the characteristics of the execution platform), if we have, for instance, the possibility of reducing some computation constraints (e.g., by increasing memory). We also might have more control over the *Personnel* and *Project* factors.

The remainder of this chapter focuses, in particular, on these two last aspects and describes what we can do to make development and the quality of the delivered product more predictable. We do so by looking, in the next section, at the way in which complex systems entail a specific development process and, then, by presenting the activities that characterize the development of critical applications.

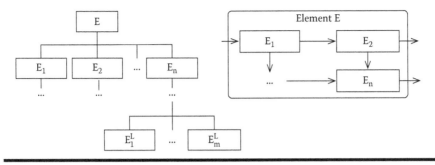

Figure 4.2   Structure of a complex system.

4.5 From System to Process

Top-down decomposition of a system into more elementary components is an effective approach to managing the complexity of system development. The decomposition can be represented as a tree, like that shown on the left-hand side of Figure 4.2. For instance, in the figure, the relationship between the element E at one level of the hierarchy and its children $E_1, E_2, \ldots, E_n$ is that $E_1, E_2, \ldots, E_n$ are all and the only elements necessary for specifying or building E. The number of levels of the hierarchy depends on the designers' choice and the complexity of the system to be developed. The leaves of the hierarchy ($E_1^L, \ldots, E_m^L$, in the figure) are components that are simple and detailed enough to be implemented.

Different approaches can be used to define and organize such hierarchies. Very common is the distinction between **function** and **product trees**. They correspond to strategies in which the system is decomposed, respectively, according to the function it must perform (function tree) or the components it is made of (product tree)—see, for example, European Space Agency (2008)[1]. Notice, however, that the strategies mentioned above are not always strictly followed and **mixed decompositions**, that is, including both functional and product characteristics, are also quite widely used.

As an example, consider the hierarchy of Figure 4.3, taken from the Department of Defense (1998), that presents a possible first-level decomposition of a military aircraft system. The advantages of the hierarchical decomposition of a system are quite wide, including the possibility of organizing the work to be performed to build a system. Each leaf of the hierarchy, in fact, represents an activity (work package) that can proceed in parallel to the other activities necessary for the implementation of other system elements at the same level of the hierarchy. Notice that when used as a project planning tool, the hierarchy is typically "enriched" with new subtrees representing the other activities that need to be performed to deliver a system (e.g., training, project monitoring). Such hierarchies come under the name of **Work Breakdown**

---

[1] Notice that the document also describes a "Specification Tree," that is, a "hierarchical relationship of all the technical specifications of the different elements of the system."

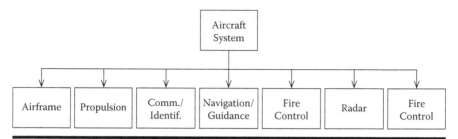

Figure 4.3   First-level decomposition of a military aircraft system.

Structures (WBSs) and are extensively used. See, for example, the Department of Defense (1998) for some guidelines in defining and using WBSs. In parallel with the construction of the product of function tree design, engineers define the system's architecture. That is, for each level of the hierarchical decomposition, a diagram is built showing how the lower elements of the hierarchy are interconnected to implement that at the higher level. This is shown on the right-hand side of Figure 4.2, which shows how elements $E_1, \ldots, E_n$ implement E.

From the point of view of system development and safety assessment, hierarchical decomposition enables two activities that greatly simplify the top-down development of a system:

1. **Top-down allocation of (safety) requirements** allows us to decompose and allocate safety requirements, according to the *obligations and benefits* approach.
2. **Early assessment** allows us to reason about system's safety on abstract representations, that are then progressively refined using *abstraction and refinement.*

4.5.1 Obligations and Benefits

The term "obligations and benefits" is taken from a software development paradigm, "design by contract," developed in the object-oriented domain in the late 1980s (Meyer, 1992, 2009). According to this paradigm, interaction among software components is based on what a component must guarantee (the obligations) and what it can take for granted about the interaction with other components (the benefits). This allows us to eliminate problems related to interfacing components. The analysis required for the allocation of benefits and obligations, in fact, helps ensure that all the necessary checks and controls are performed by one of the components.

An example might clarify this approach. Consider a simplified Anti-locking Brake System (ABS) composed of two modules. The first module is responsible for continuously reading the speed of the wheels; the second is responsible for using the data produced by the first module to actuate or release the brakes on the wheels moving significantly faster or slower than the others. Checks must be performed on the data read to ensure, for instance, that one has properly dealt with overflows and other kinds of sensing errors.

Design by contract ensures that these checks are clearly allocated as an obligation of one of the two components; the other has the benefit of assuming that the data is according to specifications. In allocating the obligation, as an added benefit, we also get a better comprehension of the interfaces among the components, including, for instance, what it means for the data to be "according to specifications."

Although the approach just described seems obvious, its application in complex projects is not always that simple, as the Mars Climate Orbiter accident has demonstrated.

On 23 September 1999, due to a navigation error, the Mars Climate Orbiter approached the red planet at an altitude of 60 kilometers, rather than the 150 kilometers initially planned. The trajectory error caused the spacecraft either to bounce on the atmosphere and be lost in outer space or be destroyed by excessive friction and mechanical stress. Post-accident investigation determined that the root cause of the accident was the use of different measurement units in two software components of the control system. One component used and assumed it was receiving values in English units, while the other adopted the metric system. The control system of the spacecraft thus underestimated the effect of the thrusters by 4.5 times, that is, the conversion factor between the two measurement systems. The spacecraft steadily deviated from the course initially planned and ended up approaching the planet with the wrong trajectory. The mission cost was estimated at $327 million (NASA, 1999).

Complex as its application might be in practice, design by contract is extremely effective for the allocation of functional and safety requirements to the elements of the hierarchy. During design, in fact, the functional and safety requirements of a system's element E are decomposed into smaller, more manageable requirements, whose implementation is delegated to E's children, $E_1, \ldots, E_n$. That is, obligations are allocated to components as we move down the hierarchical decomposition. In contrast, when $E_1, \ldots, E_n$ are integrated to implement E, the fact that E actually implements the functions and safety requirements initially allocated to it is based on the functions and safety requirements actually implemented by the lower-level components $E_1, \ldots, E_n$. That is, benefits are exploited as we move up the hierarchical decomposition.

The advantages occur both during design and verification:

■ During *design*, safety assessment of a system can proceed in a top-down fashion. Safety analysis can be performed on any given element of the hierarchy by making assumptions about the safety provided by the components at the lower level in the hierarchical decomposition. This allows us to validate decomposition and architectural choices and to allocate safety requirements to the lower levels of the hierarchy.

■ During *verification*, the flow is reversed. Assessment activities now have the goal of demonstrating that each element correctly implements the safety requirements allocated to it. This allows us to verify the hypotheses made during design and thus validate the design of higher-level components.

4.5.2 Early Assessment

The allocation of the requirements described above requires us to perform an initial assessment on the system, evaluate architectural choices, and decide what components are allocated which requirements.

Quite often such analyses are performed by hand and, consequently, are limited to simple scenarios. For instance, fault models are "injected", one at a time, in each different component of the system under analysis, to assess the possible consequences of permanent faults in a single component.

The progressive refinement entailed by the hierarchical decomposition allows us to use formal methods to perform (more) complex analyses. The trick is to use formal notations (e.g., state-machines) to specify the behavior of components. Because the specification can be rather detailed and tool supported, simulations and analyses are no longer limited to simple cases, like the scenario described above.

For example, permanent and intermittent failures and the delivery of wrong values are behaviors that can all be modeled and analyzed. As in the previous case, results can then be used to validate or revise architectural choices and continue with the development. This is shown in Figure 4.4. The top left part shows the architectural diagram of a given element of the hierarchy, which we call S. For each element of S, a behavioral specification is provided in the form of a state-machine, as shown in the upper right part of the figure for Subsystem 4. Notice that the component Subsystem 4, in turn, could be composed of various elements and its actual behavior (i.e., when is fully implemented) will be more complex than the one used for the analyses at the higher level, as shown in the lower part of the figure.

To understand what information and guidance we can obtain from this kind of early assessment, it is necessary to analyze the relationship existing between the higher-level (or abstract) characterization and its actual (or ground) implementation. To do so, we briefly recap some results about abstractions of formal representations.

One of the first theories of abstraction in theorem proving was proposed by Giunchiglia and Walsh (1992). The theory defines an abstraction as a functional mapping between two formal theories and distinguishes among **increasing** (TI), **constant** (TC), and **decreasing** (TD) abstractions, according to whether

- Results in the concrete space hold also in the abstract space (TI)
- Results in the abstract space hold if, and only if, they hold also in the concrete space (TC)
- Results in the abstract space hold in the concrete space (TD)

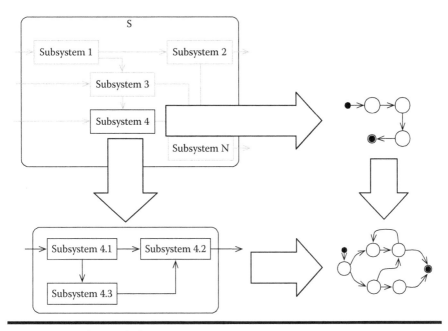

Figure 4.4   Behavioral abstraction.

Abstractions are used as a heuristic for simplifying the proof of theorems—see, for example, Giunchiglia et al. (1999). The approach can be described as a *divide and conquer* technique: theorems are first proved in the abstract space, where they are simpler to demonstrate. The proofs in the abstract space are then used to drive the proof in the concrete space, for example, by suggesting lemmas.

In the case of theorem proving, TI abstractions appear to be the most useful. As a matter of fact, although TC abstraction guarantees that each result found in the abstract space holds also for the concrete, more detailed, space (and vice versa), they are very difficult to define. TD abstractions make things even more complex because there are less theorems in the abstract space than there are in the ground space.

A similar approach has been proposed by Clarke, Grumberg, and Long, for model checking (Clarke et al., 1992, 1994). Simplifying quite a bit, abstraction is defined as a mapping between two executable representations. The goal is also similar, that is, using results in the abstract representation to prove properties about the concrete system. One important result in Clarke's theory is the characterization of the abstractions and problems for which results in the abstract representation are guaranteed to hold in the ground representation.

Thus, if we go back to our original problem, the theories of abstraction provide a clue as to the validity of the results obtained during early assessment. For instance, if we know that the model we use will necessarily result into an abstraction of the actual implementation satisfying Clarke's condition, then we are guaranteed that any result found during early assessment will necessarily hold also for the actual implementation.

4.6 A General Development Framework

Many structured processes and frameworks have been proposed to control the sources of complexity and uncertainty that characterize system development. Some emphasize the importance of standardizing products and activities. Others privilege risk mitigation and flexibility in the development; and some others highlight the importance of tools. Common to all, however, is a strict integration of the diverse disciplines and competencies that are required for building a system.

Figure 4.5 shows the main phases, workflows, and activities characterizing system development. The diagram is loosely inspired to the PMBOK (Project Management Institute, 2004), which provides a similar view for project management, to the Rational Unified Process (Kruchten, 2000; Royce, 1998), which has an analogous schema for software development, and to FAA (2000), which has a similar diagram for the development of safety-critical applications.

Time flows from left to right. On the horizontal axis, therefore, we have the different stages that characterize the development of a system, from concept to deployment. On the vertical axis we have the main workflows that concur to the successful development of a system.

There are two important characteristics that we need to highlight:

1. Together with technical workflows, such as "Development" and "Safety," we present two managerial workflows, namely, "Project Management" and "Certification." Project management activities, in fact, have a decisive impact on the quality of the final product. A similar consideration can be made about the certification workflow, because it influences technical activities and it must be planned from the early stages of product development.
2. The diagram can be replicated at different levels of detail and granularity. That is, the hierarchical decomposition of the system allows us to break down the complexity of a system by organizing its development in a **program** of interconnected projects. Each project can then be carried out according to the framework of Figure 4.5. This is shown in Figure 4.6, which is inspired by "The Systems Engineering Engine" of NASA (2007), and shows the organization of work for an element of hierarchical decomposition and its connection with the levels immediately above and below.

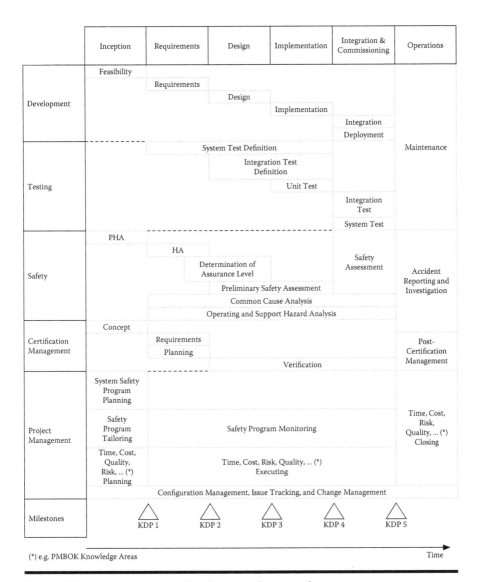

Figure 4.5   A general system development framework.

There are two different ways of looking at the diagram in Figure 4.6. By row, we get a view of the activities carried out within each workflow. Each row focuses on a specific aspect of system development. By column we see how development progresses and how the different workflows contribute to building a unified view of the system. We detail both in the following sections.

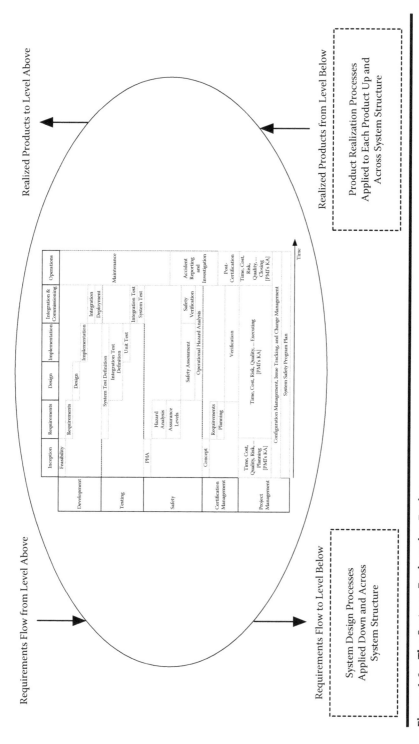

Figure 4.6 The Systems Engineering Engine.

4.6.1 Phases and Phase Transition

Transition from concept to product proceeds through the following phases:

1. **Inception.** During this phase the need (or opportunity) for a new system or for revising an existing system is drafted, elaborated, and evaluated. It is here that initial considerations regarding technical and economical feasibility, risks and opportunities, forecast profit and losses concur to determine whether or not to move to the next phases.

2. **Requirements.** During this phase the characteristics of the system are defined. An analysis of the system's objectives helps define the main characteristics of the system to be developed. In parallel, the analysis of the environment in which the system will operate, previous experience with similar systems, and experts' judgments are used to individuate potential hazards and first safety requirements.

3. **Design.** During this phase the system's structure and design are drafted and evaluated. Safety-critical components are individuated. Design alternatives (e.g., fault avoidance rather than fault tolerance) are evaluated. Technological choices are made for the development of critical and noncritical components. Activities to address the certification of the final system are also carried out.

4. **Implementation.** During this phase the system is developed according to the structure and design defined during design. The architecture defines the main components of the system, whose development and testing can proceed in sequence or parallel, according to the constraints and degrees of freedom allowed by the design and by the project. Unit testing and safety assessment ensure that the quality and safety goals of each element of the design are actually met.

5. **Integration and commissioning.** During this phase the components of the architecture are integrated into the final assembly. System test and safety assessment activities are meant to ensure that the quality, functional, and safety goals are actually met by the final system. The system is certified and put into operation. Moreover, the activities related to installation, training of resources, acceptance testing, and certification are taken care of in this phase.

6. **Operations.** During this phase the system is maintained and serviced to ensure it remains functional. Moreover, as market changes and technology evolves, some functions, parts, or components might be upgraded to improve some of the characteristics of the system.

Transition from one phase to the next is determined by the achievement of one or more milestones, marked in Figure 4.5, with triangles labeled KDP_n (Key Decision Point). Milestones are based on the analysis of the outputs of a phase. Verification activities, such as inspections, audits, and checklists, are performed to ensure that the minimum set of information and the minimum quality standards required by each milestone have been achieved.

Results of verification activities are used with two complementary goals:

1. They can be used as a formal verification point to *decide whether to transition to the next phase*. If the milestone assessment, for instance, highlights outputs below the minimum quality standards, the transition to the next phase might be delayed to fix problems.
2. They are a formal verification point to *decide whether to continue with the project*. The output of the feasibility phase, for instance, might highlight technical difficulties deemed too difficult or too costly to overcome. Problems during development might highlight intrinsic flaws in design that could make deployment of the solution too risky.

4.6.2 Comparison with Other Frameworks

The organization of project activities in phases is quite common, especially for larger and safety-critical applications. The number of phases, the level of detail, and the transition rules from one phase to the next, however, can be quite different, according to application domain and the applicable standards.

As an example, we briefly present, in the remainder of this section, phases and milestones of two different development standards: the Rational Unified Process (RUP) and the European Space Agency (ESA) standards for project planning.

4.6.2.1 The Rational Unified Process

The RUP, a software development process, organizes development into four different phases: namely, Inception, Elaboration, Construction, and Transition. Activities within each cycle can, in turn, be organized in cycles according to the complexity of the system to be developed (Kruchten, 2000). Four milestones define the transition from one phase to the next:

1. **Life-Cycle Objectives,** which concludes the Inception phase and provides stakeholders with the information necessary to decide whether or not to continue to the next phase.
2. **Life-Cycle Architecture,** which concludes the Elaboration phase and provides a stable version of vision, requirements, architectures, and prototypes demonstrating the technical feasibility of the major risk factors.
3. **Initial Operational Capability,** which concludes the Construction phase that ends up with a system ready to be deployed and with all the other activities necessary for system's operation (e.g., training) in place.
4. **Product Release,** which concludes the Transition phase by verifying that the system has achieved the initial goals and met the stakeholders' expectations.

4.6.2.2 ESA Standards

The project planning standard defined by the European Space Agency is meant to capture the entire life cycle of hardware/software systems. According to the standard, development is organized in seven different phases:

- **Phase 0, Mission analysis and needs identification.** During this phase goals and needs are characterized; preliminary studies are conducted to evaluate market opportunities, costs, and main risks.
- **Phase A, Feasibility.** During this phase the main project plans are outlined, the technical feasibility of the project is evaluated, and outputs of the previous phase are refined to get a more detailed view of the system.
- **Phase B, Preliminary Definition.** During this phase the plans defined during the previous phase are refined, technical solutions are committed to, pre-development work is started on the most critical technologies, and safety assessment activities are initiated.
- **Phase C, Detailed Definition.** During this phase the system is developed. Detailed designs are provided, engineering models are built and tested, critical components are developed and prequalified, and assembly, test, and qualification plans are developed.
- **Phase D, Qualification and Production.** During this phase the system is produced and qualified for deployment.
- **Phase E, Operations/Utilization.** During this phase the system is deployed and used. This typically includes, for ESA, the execution of all the ground operations before, during, and after launch; launch and conduct of the mission's goal (e.g., scientific experiment); and finalization of the disposal plan.
- **Phase F, Disposal.** During this phase the disposal plan, drafted in Phase B and refined up to Phase E, is executed.

Transition from one phase to the next is regulated by reviews, whose type and number depend on the specific phase. They include:

- Phase 0 has one review, **MDR, Mission Definition Review,** which includes the verification of the release of the mission statement and the assessment of the initial technical and programmatic aspects.
- Phase A has one review, **PRR, Preliminary Requirements Review,** whose primary goals include the release of preliminary plans and technical specifications, confirmation of project feasibility, and selection of technologies for the next phase.
- Phase B has two reviews:
  - **SRR, System Requirements Review,** whose primary goals include the release of the final technical specifications, assessment of preliminary design, and assessment of the preliminary verification program.

- **PDR, Preliminary Design Review,** whose primary goal is verification of the preliminary design against the project requirements. Moreover, the release of product tree, WBS, and project assurance and verification plans is verified.
- Phase C has one review, **CDR, Critical Design Review,** whose primary goals include the release of the final design, assembly, and test planning; and verification of the compatibility with external interfaces.
- Phase D has three reviews:
  - **QR, Qualification Review,** which is held during the phase and whose primary goal is ensuring that verification and validation activities are complete and accurate.
  - **AR, Acceptance Review,** which is held at the end of the phase and has goals similar to the Qualification Review.
  - **ORR, Operational Readiness Review,** meant to verify readiness for operations (the next phase).
- Phase E has four reviews. They include **FRR, Flight Readiness Review; LRR, Launch Readiness Review; CRR, Commissioning Result Review;** and **ELR, End-of-Life Review,** conducted at the end of the mission.
- Finally, Phase F has one review, **MCR, Mission Close-out Review,** meant to ensure proper disposal of the system.

See European Space Agency (2008) for more details.

4.6.3 Organization and Sequencing of Phases

The strict sequencing of phases discussed above comes under the name of **waterfall** development. It was first proposed for software development by Walter Royce, who introduced it in the 1970s (Royce, 1970). For a wide variety of projects, the waterfall is an effective—often the only appropriate way—to manage project complexity. (Notice that even when the waterfall approach is chosen, activities within each phase might be organized in more flexible ways, for example, in cycles.)

There are, however, various systems and domains for which the waterfall approach imposes a structure that is too rigid with respect to the system's objectives and complexity. This is particularly true in software development. As a matter of fact, some of the critiques that can be assigned to the waterfall approach include:

- **Efficiency.** The strict sequencing of activities often does not allow us to exploit the fact that the development of certain (sub)systems might proceed independently and therefore in parallel, possibly shortening the time it takes to deliver a system, if enough resources are available.
- **Rigidity.** As the features of the system become clearer, opportunities arise and constraints become clearer. The strict sequencing imposed by the waterfall model, because it is based on freezing choices, often does not allow us to

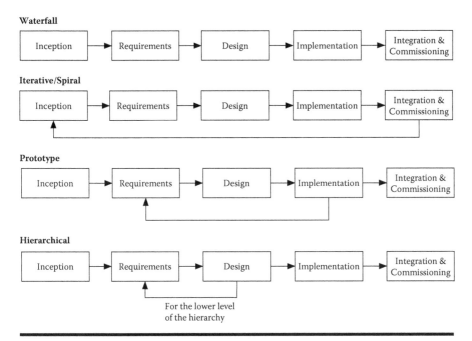

Figure 4.7   Development processes.

exploit such opportunities if they require backtracking some already-taken decision. This, in turn, tends to favor the development of clumsier solutions.

■ **Difficulty in managing risks.** In the strict sequential model, components are assembled together only at the very end. The evaluation of the system's characteristics, therefore, can be performed only at a late stage of the development process, when the cost of changes is high and the time left before delivering short.

Different paradigms have been proposed in the past that rearrange the activities mentioned above in more flexible ways. They are shown in Figure 4.7 and vary from the waterfall model by . . .

> . . . **Allowing for the construction of one or more prototypes.** In the **prototype**-based approach, the design and development phases are repeated two or more times at different levels of "abstraction." A first cycle allows to build a simplified prototype to explore new concepts or complex engineering problems. The information gotten by developing and testing the prototype is then used to drive the development of the actual system. The process is particularly suited when new technologies or new ideas are to be assessed and possibly integrated in the solution. Disadvantages include increased costs and delivery time.
>
> . . . **Allowing for an iterative development of the system.** Especially suited for software systems, in the **iterative** development process, all the phases are

repeated at various times to deliver increasingly refined versions of the system. The process allows us to focus on the most critical user functions first, and it typically shortens the time required to deliver a system that the customer can actually use. On the other hand, as iterations increase, so might problems in integrating new functions in an architecture that may have not been thought to accommodate them in the first place. A specific implementation, in which risks are analyzed first, is due to Barry Boehm and is called the Spiral process (Boehm, 1988).

... **Allowing for the parallel development of independent components.** After inception and general system requirements, the development process is organized according to the hierarchical decomposition of the system to be built, similar to that described in Figure 4.6.

4.6.4 Workflows

Effective system development requires integrating different types of activities and competencies. These are represented by the rows in Figure 4.5, where we distinguish, in particular,

1. **Development.** This workflow includes all the activities necessary to build a system. It consists of activities such as the definition of the (functional) requirements, the design of the architecture, and the integration of components. Simplifying a bit, the main goal and concern of this workflow is defining system functions and system characteristics, and implementing them.

2. **Testing.** This workflow comprises all the activities necessary to confirm that the system performs as expected. This includes verifying that the system implements all and only the functions documented in the requirements documents and that the system meets all the nonfunctional characteristics foreseen and agreed upon (e.g., performance, portability). Thus, the main focus of this workflow is to ensure that the system behaves as initially imagined and agreed upon.

3. **Safety.** This workflow collects all the activities to ensure that the system behaves as expected under degraded conditions, that is, when components fail. The activities in this workflow are strictly interrelated with those in the development and certification workflows. In fact, they contribute to the definition of additional requirements for the system (i.e., the safety requirements) and impose constraints on the design to implement them (e.g., the use of a TMR in a critical component). Integration with the certification workflow ensures that the system will actually meet safety standards and regulations required for certification.

4. **Certification management.** This workflow consists of all the activities to ensure that all the conditions for certification are met. Conditions include compliance with regulations and adherence to standards.

5. **Project management.** This workflow includes all the activities to coordinate efforts in the other workflows and achieve the project's goals, and includes aspects such as planning safety and development effort, forecasting schedule and costs, and monitoring technical achievements. An important aspect, often overlooked, is ensuring a proper flow of information among the project team and stakeholders, as this is often a key factor in ensuring that functional and nonfunctional goals are actually met.

Each workflow is, in turn, composed of more elementary activities. We describe them in the following sections.

4.7 Development Workflow

Figure 4.8 shows the activities carried out during system development. The notation used is derived from the activity diagrams of the UML. Rounded rectangles represent activities, rectangles represent artifacts, a black dot represents the initial state, and a circled dot represents the final state. Different from the standard notation, we show the content of some artifacts and we present the diagram in a grid, to highlight workflows and phases[2]. See Pilone and Pitman (2005) and Fowler and Scott (2000) for a short introduction to the UML, and Booch et al. (2005) for a more complete reference.

The workflow is presented for the waterfall development process, although activities can be rearranged, according to the guidelines given in Section 4.6.3, to support different kinds of development processes.

We distinguish the following activities:

- Feasibility study
- Requirements analysis
- Design
- Implementation
- Integration

4.7.1 Feasibility Study

The goal of the **feasibility study** is to assess the technical feasibility of the system to be developed. Preliminary discussions and analyses allow us to draft the concept of the system to be built and provide a first high-level definition of its functions and main characteristics.

---

[2] People familiar with the UML will notice some similarity between the vertical separations and swimlanes. However, different from what the UML prescribes, the vertical separation lines represent different phases in time, rather than different responsibilities.

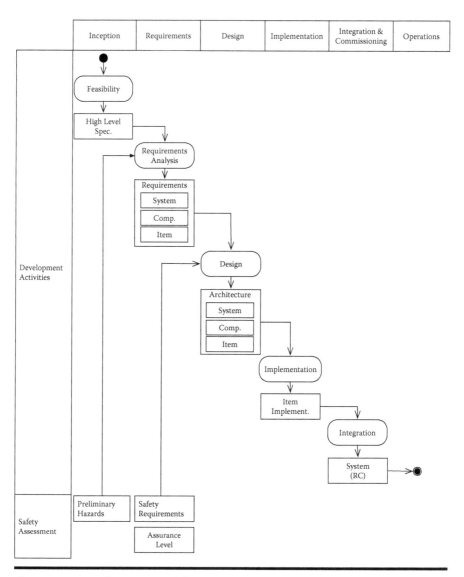

Figure 4.8   Development workflow.

The output of this activity is a **feasibility document**, which contains a high-level description of the system, together with its main functions and distinguishing features, such as the main technical innovations. The feasibility study, together with a preliminary assessment of the risks (see Section 4.9), is the information used to decide whether or not to continue on with the project, and it is the basis for defining risk countermeasures, and mitigation plans.

Good feasibility studies have both breadth and depth. The former is needed to take into account all factors possibly influencing the development of the solution. The latter is needed to anticipate possible issues on critical components. To get a comprehensive view, the 5M model of system engineering (FAA, 2000) and the guidelines for managing stakeholder expectations given in NASA (2007) are excellent starting points.

The 5M models helps in organizing system description and analysis by focusing on the following elements:

- **Mission.** This is the main purpose of a system and the reason that all the other elements are brought together. Two important aspects to investigate here are the *success criteria*, namely what the mission must accomplish to achieve the stakeholders' goals, and *strategic objectives*, namely how the mission fits into an organization's strategic plans and goals.
- **Man.** This is the human factor. Aspects such as risks, constraints, and training needs must be taken into account in order to evaluate their possible impact on mission success.
- **Machine.** This is the platform that will be used to achieve the mission's goals. One important aspect to investigate includes the definition of the *design drivers*, namely the technological choices that will influence system development. See Section 4.2 for a more detailed discussion of the topic.
- **Media.** This is the environment in which the system will operate. See Section 4.2.
- **Management.** This is the set of procedures, regulations necessary for the operations, maintenance, and disposal of a system.

4.7.2 Requirements Analysis

The goal of **requirements analysis** is to define the functional and nonfunctional characteristics of the system to be built. During this activity, the outputs of the feasibility study are refined to produce a sufficiently detailed and precise system description.

The output of this activity is a **requirements document** (or a set of requirements documents) that describes the characteristics of the system, in the form of a list of statements or narrative descriptions that tell what the system shall do. Requirements are often written in natural language, although semistructured notations (i.e., using both natural language and mathematical formalisms) or purely formal specifications (i.e., using exclusively mathematical notations) are sometimes used.

The requirements document is often used for contractual agreements with the project sponsor and with sub-contractors. It is a also good practice to have the document agreed upon by all the stakeholders in the project, including people responsible for the actual implementation of the requirements, in order to help highlight and anticipate possible technical issues in the implementation.

We further decompose requirements analysis into the following three tasks:

- **Requirements elicitation,** during which the characteristics of the system are elicited. Typical approaches to the elicitation of requirements include interviews with stakeholders, analysis of similar systems, and analysis of standards.
- **Requirements specification,** during which requirements are organized and annotated with meta-information to simplify their management. Meta-information are pairs *key-value* and help further characterize each requirement. Among the attributes most commonly used, we mention **author, version, difficulty** (i.e., how difficult the implementation of the requirement is), and **type** (e.g., Functional, Usability, Reliability, Performance, Supportability, and Maintainability). For certification purposes, an essential attribute is **traceability**, that is, the origin of the requirement. **Primitive** requirements originate from stakeholders or other project documentation (e.g., feasibility study). **Derived** requirements originate from other system requirements, during the refinement of the requirements. Traceability is essential, for example, to properly propagate changes and to assess the completeness of testing activities.
- **Requirements validation,** which has the goal of confirming that the requirements describe "the right system." During this activity, requirements are analyzed to highlight ambiguities, inconsistencies, deficiencies, and other problems. Different techniques can be used, among which we mention inspections and formal analysis. See Chapter 5 for more details.

Notice in Figure 4.8 that we distinguish three different types of requirements: **System Requirements, Component Requirements,** and **Item Requirements.** They correspond, using terminology taken from SAE (1996a), to increasingly detailed views of a system. Their role is clarified in Section 4.7.5, where we describe how requirements analysis and design are interleaved for the development of complex systems.

Note also in Figure 4.8 that we distinguish safety requirements from the other types of requirements—see the "Safety Requirements" box at the bottom of the figure. We do so because the elicitation and management of safety requirements are more properly explained by looking at the "Safety Assessment" workflow—see Section 4.9.

4.7.3 Design

During **design**, all requirements (including safety requirements) are used to determine the architecture of the system. The outputs of this activity are diagrams, documents, and blueprints describing the main building blocks of a system.

The specific strategies to choose the architecture of a system depend on the system type (e.g., software or hardware), the domain, and the technologies used. What is important to note here is that, during design, the hierarchy of the system is built and requirements are allocated to the different components of the architecture. That is, each component of the architecture becomes responsible for the implementation of one or more requirements.

The allocation of safety requirements, in particular, has two consequences on the design:

1. It imposes specific design choices, such as, for example, the adoption of a TMR for a critical component.
2. It imposes constraints on the development process that must be adopted for the critical components of the architecture. This is formally done by associating an *assurance level* to each component, depending on the safety requirements it is responsible for implementing. We look in more detail at the allocation of assurance levels when we present the "Safety Assessment" workflow in Section 4.9.

4.7.4 Implementation and Integration

During **implementation**, the leaf elements of the architecture are actually built according to their specification, following organizational standards and conventions. During **integration** the various components are assembled and integrated. The output of this activity is the "machine," using 5M terminology, ready to be tested and validated.

4.7.5 Hierarchical Design of Systems

Of particular interest is the form the development workflow takes when we adopt the hierarchical decomposition process. This is depicted in Figure 4.9, which shows the process for a system whose architecture is based on three levels: system, component, and item. Time flows from top to bottom.

The key concept is that development becomes an alternation of requirements analysis and design, conducted at different levels of detail:

■ At the first level of detail (the system level), requirements analysis individuates the system's requirements. These are used, during design, to define the system architecture and, correspondingly, the first level of hierarchical decomposition. (See the upper part of Figure 4.9.) In parallel, traceability between components and requirements is elicited and kept.

■ At the second level of detail (the component level), the system architecture, the traceability matrix, and the system requirements are decomposed, refined, and allocated to the components individuated in the previous step. This is shown in the central part of Figure 4.9. A new design phase can now define the architecture of each component, in terms of items, and allocate requirements to them.

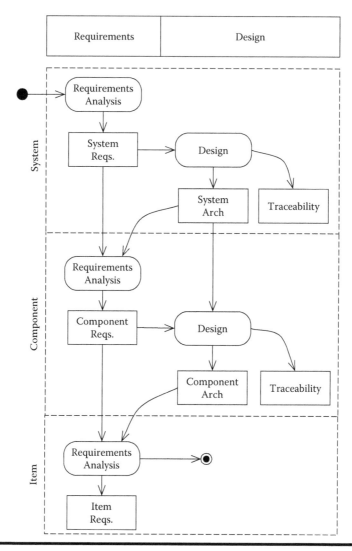

Figure 4.9   Unfolding of the development workflow with hierarchical systems.

■ At the third level of detail (the item level), if we assume items to be simple enough so as not to require a specific architectural decomposition activity, the item requirements are obtained by refining the component requirements, similar to what we did in the previous step. These, in turn, are the basis for the items' implementation.

4.8 Testing Workflow

In general, testing activities span throughout the system development activities and may continue in the production phase. We distinguish the following broad categories of testing:

- **Development testing,** which is performed during system development.
- **Verification testing,** which concentrates on the verification of the final product.
- **Production testing,** which is carried out to ensure that every item produced according to the development process does not have any fabrication defect (e.g., production testing for chips in hardware design).

Testing may have different objectives, depending on the phase in which it is carried out. During development, the goal of testing may be to uncover as many faults as possible. During the verification phase, testing may be performed to ensure that the system does not contain any faults and that it conforms to the requirements specification.

Independently from scope, testing activities are typically structured in three stages:

- A **planning stage,** during which the scope, the plan, and the general objectives of the testing campaign are drafted. The test plan typically follows the development process of the particular application; for instance, progressive testing may be incorporated to more closely match the development schedule.
- A **definition phase,** during which the testing strategy is defined and the test cases are constructed.
- An **execution phase,** during which the test campaign is conducted and results are reported. Test execution can be automated according to the system to be tested and the nature of the tests to be conducted.

In this section we focus on the definition and execution phases of development and verification testing, whose activities are shown in Figure 4.10.

Test definition activities proceed in parallel with the development and define the tests to be executed on the system, at different levels of granularity. They include

- Acceptance test definition
- Integration test definition
- Unit test definition

4.8.1 Acceptance Test Definition

The goal of **acceptance test definition** is to specify the tests to be performed on the system. The output is an **acceptance test document**, describing the tests to be executed to verify compliance of the system with the requirements.

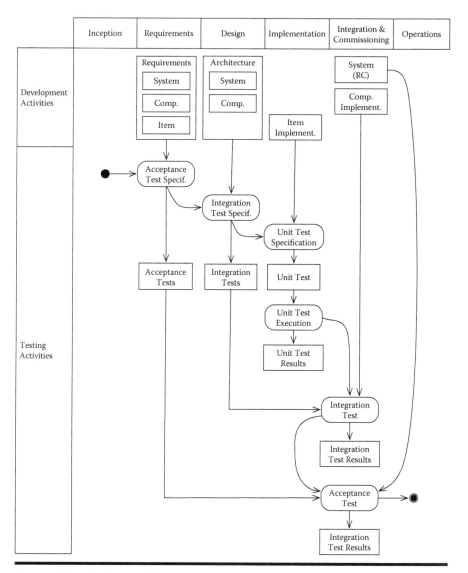

Figure 4.10   Testing workflow.

Tests are often put in the form of **test cases**, that is, semistructured specifications of the activities that must be carried out to verify the correctness of the system. Test cases can be accompanied by a specification of the method, tools, and environment needed to perform tests. For software development this could also include scripts and programs to automate software testing activities.

The relationship between test cases and requirements is, in general, many-to-many. One or more test cases are defined to test one or more requirements. For certification purposes and, more generally, as a good development practice, is it important to maintain the traceability between requirements and test cases and ensure that each requirement is tested at least once. Tests can be performed in different ways, according to the development stage and due to the different nature of a system's components (e.g., hardware, software). We mention those listed in NASA (2004):

- **Analysis,** that is, the use of mathematical modeling and analytical techniques to predict the suitability of a design, based on calculated data.
- **Demonstration,** that is, providing a means to demonstrate that a requirement is satisfied, without a systematic collection of data. An example is demonstrating that all commands of a cockpit are reachable, by having a pilot actually sit in a mock-up of the cockpit and see whether all commands can be reached (NASA, 2004).
- **Inspection,** that is, the visual examination of a realized end product. This is typically used to verify physical characteristics of components (e.g., that a component is painted in the color specified by the requirements) or conventions adopted for source code (e.g., that variables follow a specific naming convention).
- **Test,** that is, the use of an end product to obtain detailed data needed to verify compliance with the requirements.

4.8.2 Integration Test Definition

The goal of **integration test definition** is to specify the tests to ensure that the subsystems interact as expected when integrated. The output is a **test integration document** containing the test cases to be executed to show compliance.

The tests are meant to verify that all functional, performance, and reliability requirements are actually met when items and components are assembled. If we go back to the Mars Climate Orbiter example, a successful integration test should have highlighted that the two modules of the control logic used two different measurement units.

4.8.3 Unit Test Definition

The goal of **unit test definition** is to produce the tests to verify a module or item has been correctly implemented. In software development, unit tests typically take the form of (small) programs and scripts, called **unit tests**, that exercise the component being developed.

Unit tests are typically defined by looking at the actual implementation of the component and, often, they are performed by the same person in charge of development. For this reason, unit tests are often considered an *internal* activity, part of the development workflow. (Actual and official compliance of the system with requirements is performed with integration and system testing.)

4.8.4 Test Execution

Test execution activities require the delivery of the actual implemented items, components, and system. They include

- **Unit test execution.** The unit tests defined during the "unit test specification" activity are executed and their results assessed. Unit test execution is often automated. This helps ensure that any change in the code that breaks previous assumptions and hypotheses about the behavior of the system is recognized and dealt with as development progresses.
- **Integration tests.** During this activity, integration tests are executed to verify that items and components interact as expected.
- **Acceptance tests.** During this activity, acceptance tests are executed to ensure that the delivered system correctly implements the requirements.

The execution of tests might cause iterations in the development process because, if one or more tests fail, the system must be "fixed." The amount of work to be redone and the depth of the loop depend on the severity of the flaw that caused the failure. In software development it is fairly common for systems to fail one or more integration or acceptance tests. To simplify traceability, conventional names are given to the versions under test. One commonly adopted notation uses

- The term **alpha release**, for early releases of a system.
- The term **beta release**, for releases that have undergone some preliminary testing and are more mature.
- The term **RC-N**, for releases that are considered nearly ready for release. **RC** stands for "Release Candidate" and **N** is a natural number used to accommodate the uncertainties of system development and bug fixing. Should a release candidate fail some tests, a new number can be used to denote a new release candidate.

Finally, it should be noted that testing increases the confidence that the system will perform as expected, but it cannot provide any definitive answer. The main sources of uncertainty derive from the actual testability of certain requirements (e.g., extreme environmental conditions in which the system will operate), from the implementation of the testing process (e.g., inadequate coverage of requirements), and from the system itself (e.g., due to the state space of software systems).

A proper quality program, good management practices, and sound engineering principles help mitigate the first two points. Formal methods techniques help addressing the third one.

4.9 Safety Assessment Workflow

Figure 4.11 shows the safety assessment workflow. Activities in this workflow can be organized in two different phases, with complementary goals:

- During system development (requirements, design, and implementation), they have a "constructive" role, by guiding design activities to ensure that the system will meet safety goals and safety regulations for certification.
- During integration and commissioning, they have a "destructive" role. The final goal is demonstrating whether the system is safe and whether it actually meets the safety requirements. To do so it is often necessary to question all design hypotheses and assumptions, trying to "break" the system in all the obvious and not-so-obvious ways.

Thus, although the techniques used in both phases are the same (e.g., fault trees, FMECAs), the purpose for which we use them is different.

A rather general safety assessment workflow is composed of the following activities:

- Preliminary Hazard Analysis (PHA)
- Hazard Analysis (HA)
- Determination of Assurance Levels
- Preliminary Safety Assessment (PSA)
- Safety Assessment (SA), including Common Cause Analysis (CCA)
- Operating and Support Hazard Analysis (O&SHA)

4.9.1 Preliminary Hazard Analysis (PHA) and Hazard Analysis (HA)

The goal of **Preliminary Hazard Analysis** (PHA) and **Hazard Analysis** (HA) is to identify the potential hazards and safety risks of the system. The difference between the two activities is the information used for the analyses:

- PHA is performed when initial information is available.
- HA is performed when more detailed and more consolidated information is available.

The analysis consists of three steps. First, hazards are elicited. Elicitation techniques include the use of checklists, brainstorming, what-if scenarios, expert judgment, consultation of regulations, and analysis of standards. Once the hazards have been identified, their effects are determined, using the techniques presented in Chapter 3. Finally, hazards are prioritized according to their probability and impact.

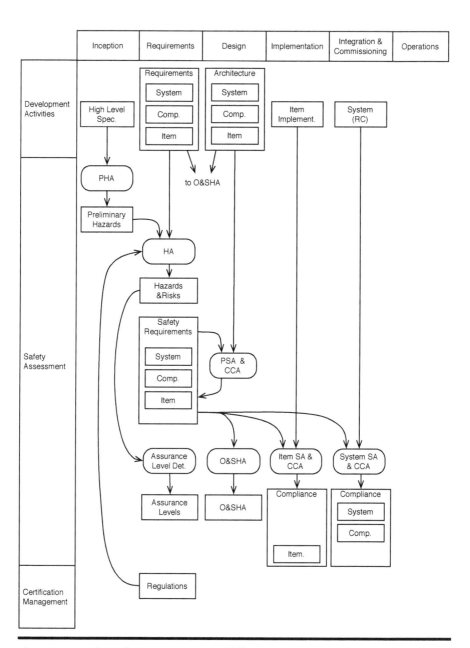

Figure 4.11   The safety assessment workflow.

Table 4.4 Developing Strategies to Manage Hazards

Description	Priority	Definition
Design for minimum risk	1	Design to eliminate risks. If the identified risk cannot be eliminated, reduce it to an acceptable level through design selection.
Incorporate safety devices	2	If identified risks cannot be eliminated through design selection, reduce the risk via the use of fixed, automatic, or other safety design features or devices. Provisions shall be made for periodic functional checks of safety devices.
Provide warning devices	3	When neither design nor safety devices can effectively eliminate identified risks or adequately reduce risk, devices shall be used to detect the condition and to produce an adequate warning signal. Warning signals and their application shall be designed to minimize the likelihood of inappropriate human reaction and response. Warning signs and placards shall be provided to alert operational and support personnel of such risks as exposure to high voltage and heavy objects.
Develop procedures and training	4	Where it is impractical to eliminate risks through design selection or specific safety and warning devices, procedures and training are used. However, concurrence of authority is usually required when procedures and training are applied to reduce risks of catastrophic, hazardous, major, or critical severity.

Source: From FAA (2000). FAA System Safety Hand book. Available at http://www. faa.gov/library/manuals/aviation/risk_management/ss_handbook/.

The output of (P)HA is a prioritized list of hazards, together with their probability and their impact. It is an input to the project management and development workflows because it determines both managerial and technical choices related to system development. Various technical documents and standards prescribe specific policies for dealing with hazards during design. One of them is the "Safety Order of Precedence," shown in Table 4.4 and taken from FAA (2000): for each hazard, the first feasible choice is taken.

4.9.2 Determination of Assurance Levels

Not all components are equally critical. We all know we can safely fly on a plane with a malfunctioning entertainment system[3]. No one, however, would be equally at ease with an engine shut down.

On the other hand, if certification standards required the entertainment system to be developed and verified using the same techniques as applied for the fly-by-wire system, development costs would skyrocket and safety would not be improved. The rigor of the development of different elements of the architecture, therefore, must be chosen according to their criticality. This is done through a process that defines and assigns **assurance levels** to each component of a system's architecture. Each assurance level defines guidelines and constraints that must be taken for developing a component.

More specifically, the assurance level is determined based on the effects that the violation of a safety requirement—for instance, due to a malfunction—could have. Typical classifications include three to five items, such as

- **Catastrophic,** if a malfunction causes unsafe operational conditions
- **Hazardous,** if a malfunction severely reduces operational safety and might cause severe stress on operators, and have adverse effects on occupants
- **Major,** if a malfunction significantly reduces safety margins and increases workload of operators, causing distress to occupants
- **Minor,** if a malfunction causes a slight reduction in safety margins, a slight increase in workload, and some inconvenience to occupants
- **Negligible,** if the requirement or component has no effect on safety

This is not enough, however, as the probability of the event is another key factor in determining how relevant a particular malfunction might be. It is a lot better to drive a car with a probability of a catastrophic failure of 1×10^{-9} per hour of operation[4], than driving one with a systematic hazardous failure.

Hazards, therefore, can be organized in a table, such as that shown in Table 4.5, where the columns classify risks (as described above) and the rows represent the probability of the event, expressed according to some qualitative or quantitative scale. The table clearly distinguishes two different areas: the upper-right part (pictured in gray) and the lower-left part (pictured in white).

The upper-right part corresponds to unacceptable design choices: No system can have a component with high probability of failure (let us say between 1 and 10^{-5} per hour of operation) and a catastrophic effect when it fails.

In qualitative risk management, three strategies are defined for risks falling in the gray area: mitigation, avoidance, and transfer. The first reduces the probability

---

[3] On the assumption that the entertainment system is isolated from the critical components of the airplane, as it is usually the case.

[4] One billion hours is slightly more than 114,115 years.

Table 4.5   Risk and Probability Matrix

	Negligible	Minor	Major	Hazardous	Catastrophic
1					
Probable					
10^{-5}					
Improbable					
10^{-9}					
Extremely Improbable					
0					

of occurrence, the second eliminates the risk (see the first policy of Table 4.4), and the third transfers the risk to a third party—for example, an insurance company. In designing safety-critical applications, only the first two options are acceptable. If a hazard occurs in the gray area, then the only design options available are either to reduce the probability of failure or to mitigate its effects. This is usually achieved, for example, by adding redundancy to the system.

Moving out of the gray area in Table 4.5, however, is not always possible or feasible, as the Concorde story teaches us. The airplane had a nearly flawless safety record until, on 25 September 2000, a Concorde was destroyed shortly after take-off.

A tire blew out during take-off and caused a rupture in a fuel tank. Leaking fuel ignited, the resulting fire compromised vital systems, and the plane eventually crashed into a hotel. Everyone on board and four people in the hotel died in the accident. After the causes of the accident were established by the final report on the accident (BEA, 2002), various design changes were proposed to avoid the risk of a similar accident recurring.

We mention the adoption of rupture-resistant tires and the insertion of Kevlar reinforcement to the tanks. Both changes were aimed at reducing the probability that the failure would occur. Decreasing demand for high-priced flights, decreasing number of passengers after 11 September 2001, and the age of the Concorde resulted in Air France and British Airways, the companies flying the plane, retiring it. Now, one Concorde is on display at the National Naval Aviation Museum in New York.

Assurance levels can be associated to areas of Table 4.5, because different zones of the table correspond to different criticality. Intuitively, as we move closer to the gray area, higher levels of assurance are needed. Because components can be associated to hazards (through the allocation of requirements to the components) and hazards to assurance levels (as we have just seen), this allows us to determine the assurance level required for each component. Two examples might help clarify the process.

The ARP4761 standard for avionic systems defines five different assurance levels, called A to E, with A being the one with the most stringent requirements (SAE, 1996b). Association of an assurance level to a component depends on the effect of

Table 4.6 ARP Assurance Levels

1		Negligible	Minor	Major	Hazardous	Catastrophic
10^{-5}	Probable	E	D	C	B	A
10^{-9}	Improbable	E	D	C	B	A
0	Extremely Improbable	E	D	C	B	A

failures, independent of the probability. Thus the table to assign an assurance level looks like that in Table 4.6.

NASA (2004) proposes instead a more articulated process for determining the assurance levels for software development. It consists of the following steps:

1. Determine the *risk index* of the software component. The risk index is determined by looking at the probability and impact of failures and could be computed, for instance, as the product of risk and impact, or by using a table, such as that shown in Figure 4.6 (adapted from NAS, 2004, page 28). The lowest risk category corresponds to an acceptable area: safety effort is optional. The highest risk category corresponds to an unacceptable area: the hazard has to be mitigated.
2. Determine the *control category* of the software component. This is done by looking at the characteristics of the software according to what is defined in MIL-882C, a military standard for development of safety-critical applications Department of Defense (1993)—see Table 4.7. (Notice that the standard has been superseded by Department of Defense (2000), where the control categories have been eliminated.)
3. Determine the *software risk index*. The software risk index is a number, from 1 to 5, corresponding to the required safety effort necessary for its development. The index is calculated using a table, shown in Table 4.8, that associates *risk index* and *control category* to a *software risk index*.

4.9.3 Preliminary Safety Assessment (PSA)

The goal of this activity is to perform safety analyses on the system as it is being developed, in order to suggest design and architectural choices to meet the safety goals. The output of this activity is a set of analyses, conducted using the techniques we presented in Chapter 3.

Of particular interest is the interaction between preliminary safety assessment and design, in the case of the hierarchical development process. This is shown in Figure 4.12, for Requirements, Design, and Preliminary Safety Assessment activities conducted on a system made up of three hierarchical levels (System, Component,

Table 4.7 MIL STD882C Software Control Categories

Software Control Category	Degree of Control
I	Software exercises autonomous control over potentially hazardous hardware systems, subsystems, or components, without the possibility of intervention to preclude the occurrence of a hazard. Failure of the software, or a failure to prevent an event, leads directly to a hazard's occurrence.
IIA	Software exercises control over potentially hazardous hardware systems, subsystems, or components allowing time for intervention by independent safety systems to mitigate the hazard. However, these systems by themselves are not considered adequate.
IIB	Software item displays information requiring immediate operator action to mitigate a hazard. Software failures will allow, or fail to prevent, the hazard's occurrence.
IIIA	Software item issues commands over potentially hazardous hardware systems, subsystems, or components requiring human action to complete the control function. There are several independent safety measures for each hazardous event.
IIIB	Software generates information of a safety-critical nature used to make safety-critical decisions. There are several redundant, independent safety measures for each hazardous event.
IV	Software does not control safety-critical hardware systems, subsystems, or components and does not provide safety-critical information.

Table 4.8 NASA Software Risk Indices

	Negligible	Moderate	Critical	Catastrophic
I	4	3	1	1
IIA & IIB	5	4	2	1
IIIA & IIIB	5	5	3	2
IV	5	5	4	3

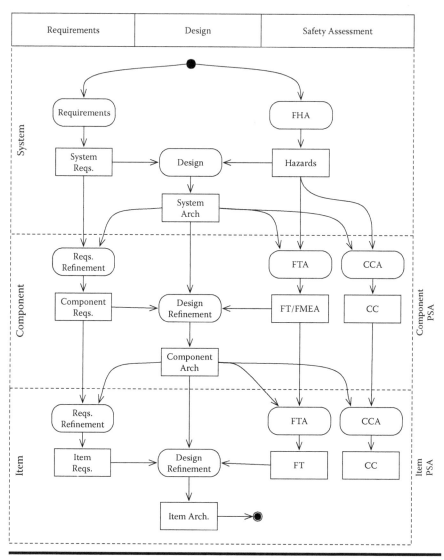

Figure 4.12   Unfolding the safety assessment workflow with hierarchical systems.

Item). The first two columns contain the Requirements and Design workflows which we have already shown in Figure 4.9. Time flows from top to bottom.

The top part of Figure 4.12 shows the activities performed at the System level. Requirements and Design (left part) define the requirements and the architecture of the system. Hazard Analysis determines the high-level safety requirements and therefore some high-level constraints for the system architecture (e.g., what elements

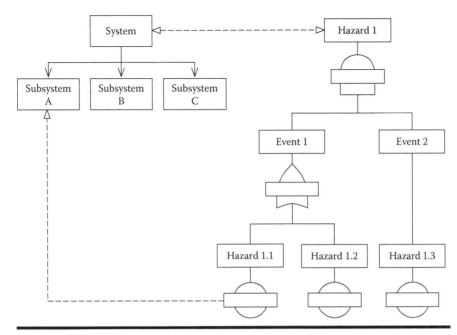

Figure 4.13   Fault trees and hierarchical development.

must be implemented using a redundant architecture). The main outputs are the requirements (allocated to elements of the architecture), the architecture, and the list of hazards (allocated to elements of the architecture). The activities proceed to the next level of refinement, the Component level. This is shown in the central part of Figure 4.12.

On the one hand, given the architecture, the system requirements are refined and allocated to the different components. This provides the specification of each component of the architecture, as already described for the development workflow. On the other hand, the Preliminary Safety Assessment performed at the component level refines the hazards into more elementary hazards that can be allocated to the different components. The refinement can be performed using Fault Tree Analysis (FTA in the figure) and is illustrated in Figure 4.13. In particular, for each hazard, a fault tree is built to decompose the hazard into lower-level contributing causes (right-hand side of figure). Such lower-level contributing causes are then allocated to the various elements in which the system has been decomposed. In parallel with fault tree analysis, Common Cause Analysis (CCA in Figure 4.12) helps ensure that the independence hypothesis holds (see Section 4.9.5). The safety requirements thus individuated are used to guide the design of the component.

The last level is that of the Items. It proceeds similarly to the component level. Activities in the development workflow refine requirements and allocate them to each item. Preliminary safety assessment activities refine safety requirements and allocate them to items.

4.9.4 Safety Assessment (SA)

As just seen, during system construction the top-down propagation of safety requirements to the lower levels of the architecture ensures that the system safety goals are actually met, *on the hypothesis that the components, once implemented, will actually meet the safety requirements they have been allocated.*

During system verification, safety assessment activities proceed in bottom-up fashion, from the lower-level components up to the system level, and are performed to verify that the hypotheses made during design are actually met. This allows us to "propagate" the results from the lower levels of the architecture to higher levels.

The safety verification activities, in particular, include

- **Item and Component Safety Assessment.** During this activity, the lower levels of the architecture are verified to ensure that they comply with the safety requirements. Fault Tree Analysis, FME(C)A, and Common Cause Analysis are used to verify compliance.
- **System Safety Assessment.** During this activity, the architecture is verified to ensure that it complies with the safety requirements. Fault Tree Analyses, FME(C)A, and Common Cause Analysis are used to verify compliance.

4.9.5 Common Cause Analysis (CCA)

On 4 June 1996, an unmanned Ariane 5 rocket just launched from the Kourou base was destroyed about 40 seconds after take-off when it veered off its flight path. Its cargo, valued at about $500 million was lost in the accident.

Investigation after the accident determined that the failure was due to an overflow caused by an arithmetic conversion performed by a software component of the Inertial Reference System. (The Inertial Reference System feeds the on-board computer [OBC] of the Ariane with information about position, orientation, and speed.) The software component causing the accident had originally been developed for the Ariane 4 and reused without changes in the Ariane 5. The Ariane 5, however, produced significantly higher values for some of the variables used by the component (Lann, 1997; Report by the Inquiry Board, 1996).

The Inertial Reference System (SRI) is implemented with a redundant architecture, based on hot-standby. However, both the primary and the backup systems used the same hardware and software. Because the problem was in the software, both components failed in the same way. (Actually, the hot-standby failed 72 milliseconds before the primary system did.) After failure, the SRI sent a diagnostic pattern to the OBC that erroneously interpreted it as valid data. This caused the OBC to command the missile in a wrong trajectory.

To make things worse, the component that caused the exception is responsible for performing tasks only when the rocket is on the ground; the component, however, was kept running also after lift-off.

The Ariane accident is a very good example of systematic failures, which we discussed in Section 2.8.4 and makes redundant architectures useless. Failures due to common causes are an important risk factor in safety-critical systems. For instance, Radu and Mladin (2003) studied the root causes of failures of the nuclear research reactor TRIGA SSR-14 located in Romania. The data collected covers a period of 20 years, from 1979 to 2000. Fifty-four events for all components have been interpreted as common cause failures. Twenty-five of them involved pumps (46%) and sixteen control rods and control rods drives (about 29%).

Common Cause Analysis (CCA) is the activity performed to ensure that the independence hypothesis holds. There are different ways of conducting CCA. For instance, SAE (1996a) distinguishes three different kinds of activities: **Common Mode Analysis**, **Particular Risk Analysis**, and **Zonal Safety Analysis**. These are described in more details in the following paragraphs.

Particular Risk Analysis. Particular risk analysis considers external events that could invalidate the independence hypothesis by causing the failure of otherwise independent components. For the avionic sector, examples of particular risks include fire, snow, lightning strike, ruptures, leakages, and bird strikes[5].

The analysis of particular risks is based on the following steps:

1. Determine the scope of the analysis (the risk to analyze).
2. Define the models regulating the risk.
3. Define the effects of the risk, based on the model.
4. Determine the failures on the system's components (based on the effects of the risk)
5. Determine how the failures propagate to the rest of the system

Zonal Safety Analysis. Physical proximity is the second way in which the independence hypothesis might be violated. If two components are installed close to one another, hazards involving a particular zone of the system might cause otherwise independent components located nearby to fail together.

---

[5] Bird strike accidents have made the headlines with the US1549 accident, in which the pilots of US1549 managed to emergency-land on the Hudson River with both engines shut, after a collision with a flock of birds. No one was hurt. Bird strikes have been on the rise in the past 10 years. According to data reported in the *Wall Street Journal*, bird strikes have been responsible for the death of 219 people and the destruction of 200 planes and cost about $628 million a year (Conkey and Pasztor, 2009). The Federal Aviation Administration maintains an online database of bird strikes (Federal Aviation Administration, 2009; Embry-Riddle Aeronautical University, 2009). The database collects information also about accidents with other species, not only birds, including various "exotic" cases, such as accidents with armadillos and iguanas.

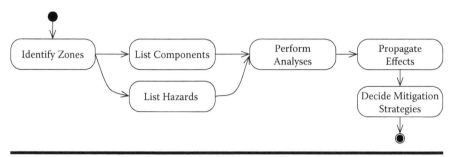

Figure 4.14   Zonal safety analysis.

Zonal Safety Analysis has the goal of identifying such kinds of dependencies. The goals of the analysis include verifying compliance with installation rules ("rules of the thumb" to minimize risks), verifying the independence hypothesis (due to system interference or particular risks), and highlighting possible problems related to errors and damages that could be done during maintenance, due to the physical proximity of components.

Zonal Safety Analysis can proceed as detailed in Figure 4.14 and consists of the following steps:

1. **Identify system's zone.** During this phase, the different zones of the system to be analyzed are identified. For systems such as airplanes, such zones are often predefined and standardized.

Then, for each zone of the system to be analyzed:

1. **List components.** All components of the zone are listed.
2. **List hazards.** All hazards pertaining to the zone are listed. Some of the hazards might derive from the risks identified in the Particular Risk Analysis (e.g., an external event); others might derive from dangerous items (e.g., combustible, pressurized components, high-voltage circuits, acids, etc.).
3. **Perform analyses.** The goal of this activity is to identify the effects of the hazards. The analysis can be conducted using FMEA or FTA, or other techniques.
4. **Define mitigation rules.** The goal of this activity is to define how to deal with the hazard due to zonal installation.

Common Mode Analysis. Common mode failure is the third way in which the independence hypothesis can be violated. Common mode failures refer to systematic errors in the development process (e.g., errors during software or hardware development) or, for example, flaws during servicing and operations.

The analysis is usually performed through the use of checklists.

Finally, we remark that the probability of common cause failures can be reduced by adopting redundant solutions based on different architectures (e.g., one hardware and one software component) or regulated by different physical phenomena (e.g., using a thrust reverser and wheel braking to slow a plane on the runway). This is, in fact, what ARP prescribes for systems in category A and is described in more detail in Chapter 6.

4.9.6 Common Cause Analysis and Software

Under the hypothesis that there are no systematic design or production errors, replication of the same hardware component—together with effective isolation, detection, and switch-out mechanisms—is a way to improve the overall reliability of a system. In such situation, in fact, failures of components might be caused by their natural wear-out, by non-systematic faults in producing components, and by external sources such as lightning strikes.

In moving our attention to software, however, things are somewhat different. O'Connor (2003) presents 13 key differences between reliability of software and hardware. We mention the following:

- Failures in software are primarily design problems.
- There are no variations: All copies of a software are identical and will fail in the same way.

Independence, therefore, cannot be achieved by replicating the same component, as errors in software derive from its production and thus are systematic. Different replicas of the same software will fail in the same way—see the Ariane example above.

One technique that is usually adopted to achieve independence of software components is called **diversity**. Different versions of the software are developed by independent teams, starting from the same specifications. The versions are often required to be implemented using different tools and programming languages. Under the hypothesis that the specifications are correct (and thus that there are no errors there), the idea behind diversity is that independent teams will not make the same design errors, and thus the different versions of the software will fail in statistically independent ways.

The independence hypothesis has been questioned by Leveson and others in Knight and Leveson (1986). The main argument is that the complex parts of a specification are the same for everyone: People will therefore tend to make the same mistakes. On top of that, programmers are taught similar techniques. Consequently, they will tend to choose similar approaches to solve problems, thereby increasing the chance of coincident errors. To prove her argument, Leveson set up experiments in which independent teams were given the same specifications and were asked to develop a system based on those specifications. The different implementations showed, in fact, coincident errors.

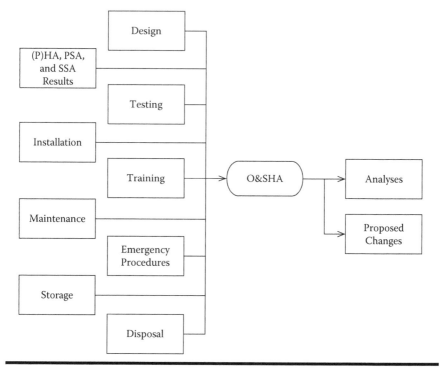

Figure 4.15   O&SHA elements.

More recent studies (e.g., see Cai et al. (2005)) performed similar experiments, leading, however, to different conclusions. Coincident failures do in fact occur, but on very rare inputs.

4.9.7 Operating and Support Hazard Analysis

The goal of **Operating and Support Hazard Analysis** (O&SHA) is identify, analyze, and evaluate the possible hazards to people, environment, and property associated to the system life cycle (FAA, 2000).

Figure 4.15, adapted from FAA (2000), shows inputs and outputs of the activity. In particular, O&SHA takes as input the information required to identify and assess hazards related to the system's production, operations, and disposal. This information comes from a variety of sources such as, for instance, design documents, installation instructions, storage instructions, emergency procedures, training, and results of other safety assessment analyses. The outputs are usually changes to design, procedures, testing, training, etc., to mitigate or eliminate hazards. O&SHA should start early in the development process, so that operating hazards are identified during the first stages of development, when changes can be accommodated more easily.

O&SHA progresses with system development activities and is usually performed in iterations, as information gets more detailed.

4.10 Certification Management Workflow

Various sectors require that safety-critical systems be certified by appropriate bodies, such as the EASA (European Aviation Safety Agency) and the NRC (Nuclear Regulatory Commission), before they can be put into operation. The certification management workflow helps to ensure that product certification proceeds as smoothly as possible. This is achieved by involving the certification authority from the early stages of development, thus anticipating and timely dealing with any possible obstacles to certification.

The workflow we present in this section is taken from AIA, GAMA, and FAA Aircraft Certification Service (2004) and FAA (2007), describing the process for the avionic sector.

The documents describe three different kinds of activities and actors that contribute to the certification process:

1. **Contractual activities,** that is, the activities to ensure that all legal aspects and agreements between the certification authority and the applicant are properly taken care of.
2. **Managerial activities,** that is, all the activities related to managing the certification process. They include planning, monitoring, and ensuring a proper integration of technical certification activities with the other workflows.
3. **Technical activities,** that is, all the activities related to, for example, evaluating compliance of the system with the target regulations and safety goals.

Certification activities progress together with product development. The AIA, GAMA, and FAA Aircraft Certification Service (2004), in particular, individuates five phases for the certification process:

1. Conceptual design phase
2. Requirements definition phase
3. Compliance planning phase
4. Implementation phase
5. Post-certification phase

Transition from one phase to the next happens similar to what was described in Section 4.6.1; audits and checklists are used to assess the achievement of the goals prescribed by the phase and their quality. The information is used to decide whether to proceed to the next phase. Let us now discuss each phase in more detail.

Conceptual design phase. The goal of this phase is *"to ensure early, value added, joint involvement with an expectation to surface critical areas and the related* *regulatory issues"* (AIA, GAMA, and FAA Aircraft Certification Service, 2004). The tasks include

- **Process orientation,** during which a partnership is established between the applicant and the certification authority. The goals include the identification of roles and responsibilities, and familiarization with product and regulations. The information is the basis for the "Partnership for Safety Plan" (PSP), an umbrella agreement between the parties on the certification project.
- **Pre-project guidance** and **familiarization briefings,** during which preliminary discussion about policies, regulations, and product characteristics allows both parties to identify potential issues, risks, and points of attention for certification.
- **Certification planning,** which outlines the certification plan, namely how the applicant will demonstrate compliance with regulations. The plan includes the data that will be submitted, methods of compliance, and a schedule of the activities.

Requirements definition phase. The goal of this phase is to ensure that the certification project is up and running. Tasks include

- **Project setup,** during which the organizational structure, team, and tools for certification are set up. This includes
  - Assigning responsibilities to the project manager and to the team
  - Forming the certification team
  - Refining the certification plan into a "Project Specific Certification Plan" (PSCP), that is, a detailed plan of certification activities
- Establishing the **issues book**. The issues book is a repository of the **issue papers**, documents that describe regulatory and technical issues related to certification, which are discussed and solved during the certification.
- Establishing the **certification basis**, that is, detailing how compliance with safety regulations will be shown.

Compliance planning phase. This is the phase during which the activities for the actual implementation of the certification program are defined. During this phase, the exact involvement of the certification authority is agreed upon between the parties.

Participation of the certification authority is required on every project decision or event that is critical for system safety. Other events are usually considered acceptable for delegation. At the end of this phase, the PSCP is updated, and it contains the detailed schedule and activities required for certification.

Implementation phase. This is the phase where the actual implementation of the certification plan takes place. Tasks include **Conformity Inspections**, which are meant to provide objective documentation of compliance of tested material with design; and **Tests**, **Inspections**, and **Analyses**, which are meant to demonstrate the actual compliance with safety regulations. See Chapter 6 for more details.

Post-certification phase. This is the phase during which the indications for managing and maintaining certification throughout the life cycle of the product are defined.

The Issues Book. We conclude this section by describing one interesting aspect covered by the document, namely how to deal with issues and disagreements between the production team and the certification body. The goal of the procedure is to properly trace and deal with all the issues, while at the same time sticking to the original development plans as much as possible.

This is achieved with the *issues book*, which collects all critical aspects of certification. Issues are added as problems are identified and arise. The *issues book* is reviewed regularly and solutions to issues are discussed and agreed to between the production team and the certification body. If no agreement can be found on an issue, this is articulated and detailed in a report, called **white paper**, stating the problem, positions, and options.

The most common type of issue paper defines particular methods of compliance with regulations. These are justified by peculiarities in the design (e.g., a new technology is used), specific conditions, or difficulties in establishing the environment under which substantiation must be shown.

Compliance in these cases can be shown using one of the following methods:

1. **Standard techniques:** The issue is shown to be properly handled using safety analysis techniques.
2. **Equivalent level of safety:** Literal compliance with the regulations does not exist, but compensating factors are put in place to deal with the issue.
3. **Special condition:** This condition is granted when the adoption of an unusual design feature or new technique does not allow us to apply standard practices.

4.11 Project Management Workflow

In general, following the Project Management Institute (2004), we define the goals of the project management workflow as follows:

> "The application of knowledge, skills, tools, and techniques to project activities to meet project requirements."

There are two mainstream frameworks dedicated to project management: the British "Project in a Controlled Environment" methodology (PRINCE2) (Office of Government Commerce, 2005) and the American "Project Management Body of Knowledge" (PMBOK) (Project Management Institute, 2004). Both are very comprehensive and include all activities, best practices, and techniques required to successfully manage a project.

In the context of the development of safety-critical applications, the goals of project management workflow widen in scope and, adapting the wording of the Department of Defense (2000), include

> "All actions taken to identify, assess, mitigate, and continuously track, control, and document environmental, safety, and health mishap risks encountered in the development, test, acquisition, use, and disposal of systems, subsystems, equipment, and facilities."

Figure 4.16 provides a simplified view of project management activities. We distinguish safety management activities (upper part) from other project management activities (lower part). Although the distinction is somewhat artificial, because

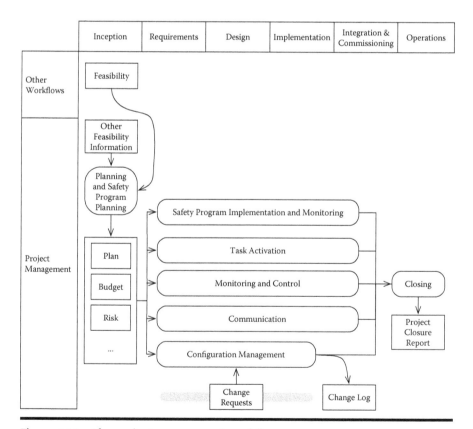

Figure 4.16 The project management workflow.

defining a sharp boundary is often impossible, it helps to further stress and emphasize the importance of a planned approach to safety activities.

The safety program management activities include

- Safety Process Definition and Tailoring
- Safety Program Implementation and Monitoring

The "generic" management activities include

- Planning
- Task Activation
- Monitoring and Control
- Communication
- Configuration Management
- Closing

4.11.1 Safety Process Definition and Tailoring

The goal of **Safety Process Definition and Tailoring** is to customize the development process to adapt it to the level of formality required to guarantee a satisfactory level of safety. This is an important activity that helps balance contrasting project requirements (e.g., time and cost on the one hand, rigor of development on the other) without compromising the safety of the resulting product.

Various guidelines for process tailoring are available. For instance, the Department of Defense (1993) distinguishes among three different types of safety critical projects—small, medium, and large—and specifies the most appropriate safety activities for each kind of project. See Appendix A of Department of Defense (1993) for more details[6].

In general, the output of this activity is a document that describes, at a minimum,

- How the safety program will be implemented, and the techniques and the standards used.
- The resources responsible for the implementation of the safety plan. This is typically achieved through the identification of a dedicated safety team and, for complex projects, a dedicated safety manager.
- How the safety program and the safety team fit into the overall project structure and the lines of communication. Because safety activities address cross-cutting concerns, the safety team is typically required to interact with experts in other disciplines and workflows.

---

[6] Notice, however, that the standard has been superseded by Department of Defense (2000), where the table does not appear.

- The hazard reporting procedures and how hazards and residual mishap risks will be tracked, communicated to, and accepted by the appropriate risk acceptance authority.

4.11.2 Safety Program Implementation and Monitoring

The goal of this activity is to ensure that the safety program is implemented according to the standards and rules defined for the project.

The first important task performed during this activity is the management of all identified hazards. Concerning this point, according to the Department of Defense (2000), the safety manager is responsible for

- Maintaining a proper log of all identified hazards and all residual hazards after development.
- Formally approving any residual hazard.
- Properly communicating to the end users all the hazards and residual hazards.

The second main task is to ensure a proper flow of communication in the project. In fact, although the ultimate responsibility for safety is with the project (program) manager, the commitment and involvement of the entire project team is essential to achieve it. These, however, are of little use if the information flow is inadequate. See, Joint Software System Safety Committee (1999) for some guidance and information on the matter.

4.11.3 Other Management Activities

Other management activities include

- **Planning.** The goal of this activity is forecasting the main dimensions that will characterize the project. Based on information about technical feasibility and risks, the output of this activity is a set of documents that defines project scope, schedule, budget, risks, communication, procurement, and quality. This set of documents serves as the baseline for the organization of all other activities of the project.
- **Task Activation.** Based on the schedule and actual project progress, the project manager activates and verifies the termination of the tasks composing the plan. Different levels of formality can be adopted, ranging from specific meetings to more informal communication among the project team.
- **Monitoring and Control.** On a regular basis, during project execution, data about the actual progress is collected and compared with the baseline plans. This allows us to forecast the longer-term impact of deviations to the plan and decide on proper mitigation actions, and update the relevant plans.

- **Communication.** Successful projects rely on an effective communication strategy that ensures that all the relevant information is delivered to the right stakeholder, at the right time. The project manager is responsible for ensuring that the information actually flows as planned.
- **Configuration Management.** Configuration management is a management discipline applied over the product's life cyle to provide visibility and control over changes (NASA, 2007). Good configuration management practices help ensure consistency between product and information and, in the words of NASA (2007), "avoids the embarrassment of stakeholder dissatisfaction and complaints." It is also an essential practice to ensure safety and for certification.
- **Closing.** Finally, when the project ends, it is good practice to assess the project, highlighting strengths and weaknesses, capitalize on the experience, and improve the probability of success of the next projects.

We conclude this brief description by emphasizing that poor project management practices are a significant cause of project failure, as highlighted, for instance, by the Mars Climate Orbiter case. When presenting the accident, we described the technical error that led to the disaster.

The accident report (NASA, 1999) goes a step further and cites the following factors as contributing to the root failure:

1. Undetected mismodeling of spacecraft velocity changes
2. Navigation team unfamiliar with spacecraft
3. Trajectory correction maneuver number 5 not performed
4. System engineering process did not adequately address transition from development to operations
5. Inadequate communications between project elements
6. Inadequate operations navigation team staffing
7. Inadequate training
8. Verification and validation process did not adequately address ground software

Notice that of the eight contributing causes mentioned, only one of them refers to a technical problem (cause number 3). All the remaining causes are, in one way or another, related to management practices that should have been in place during the project.

4.12 Tool Support

Figure 4.17 provides an overview of tools and tool-supported techniques commonly used for system development. The diagram focuses on the workflows more specifically related to development, verification, and safety assessment and is inclined toward software development. We only partially cover project management. Finally, no specific tool is associated to certification workflow. The tools used for certification, in fact, are those available for design, development, and safety assessment.

The diagram quite clearly shows that tool support varies significantly, with some activities broadly and thoroughly covered and some others in which automation is limited to the usage of word processors and spreadsheets.

In the next sections we provide a more in depth description of the diagram, listing also some of the tools freely or commercially available. We need to remark, however,

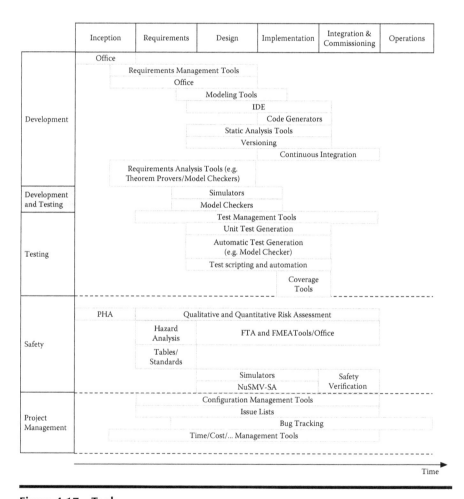

Figure 4.17   Tools.

that the listing is incomplete and must not be intended as the output of a tool-selection activity, which typically has to take into account project and organization specific requirements.

4.12.1 Supporting the Development Workflow

This is probably the most mature area with respect to tool availability. We distinguish, in particular,

- Requirements management tools
- Requirements analysis tools
- Modeling tools, simulators, and code generators
- Static analysis tools
- Versioning systems
- Build and continuous integration
- Configuration management
- Issue and bug-tracking management

Requirements Management Tools. These tools have two main tasks:

- Annotate/enrich requirements with meta-information that helps classify requirements, to simplify their management.
- Maintain traceability among requirements, in order to simplify, for example, propagation of changes.

There are basically two different approaches to the implementation of requirements management tools. Some tools are integrated in commercial word processors and allow the annotation of text with meta-information. The meta-information is stored in a database. Requirements, thus, are written and exchanged as text documents, written using familiar tools. The annotations managed by the tool allow requirement analysts for a more formal treatment of the requirements. This is the case of IBM Rational Requisite Pro, for example. Other tools take a different approach and provide their own interface, in some cases, web based. Requirements are written using the tool and exported in textual formats when documents need to be exchanged. The storage of requirements is typically based on some kind of database. Advantages include a more formal treatment of the requirements and parallel editing. This is the case, for instance, of IBM Rational Doors.

Various resources are available on the web to get more information on requirements management tools. We mention INCOSE (2009), which provides a survey of several different tools for requirements management.

Requirements Analysis Tools. These tools are used to reason about requirements and provide information about consistency and completeness. Most of these tools are based on formal representations that must be used to represent requirements. Analyses are performed interactively or automatically; the inference engines are usually based on model checkers or theorem provers.

In this area we mention methodologies such as B, Z, and tools such as RAT, which we describe in more detail in Chapter 5.

Modeling Tools, Simulators, and Code Generators. Modelers and simulators blur the distinction between design, development, and testing by allowing early assessment, simulation/verification of the design, and, in some cases, automatic generation of (certified) code. The advantages in terms of productivity and accuracy are enormous.

Tools developed for hardware systems saw an earlier development. We mention SPICE, MATLAB®, and Simulink, which allow for the design and simulation of complex mixed systems. Tool support for software development saw a steady growth in the past 20 years. Languages and associated tools, such as Lustre/SCADE, Statecharts/ Statemate, and UML+SDL/IBM Tau, allow us to specify systems at a higher level of abstraction. Tools support simulation, code generation, and round-trip engineering. Some of these tools are also integrated with theorem provers and model checkers.

Formal verification is also possible using dedicated tools and environments. Various model checkers, such as Spin, the SMV-based family of model checkers (among which SMV, Cadence SMV, and NuSMV) are available for noncommercial or free use. Finally, theorem provers, such as PVS, ACL2, HOL, and Isabelle, provide environments for the verification of requirements and to demonstrate compliance of the design with (functional) requirements.

Static Analysis Tools. Software development relies on the use of Integrated Development Environments, some of which are freely available. Certified compilers might help reduce the testing and certification burden. (See Chapter 6 for more details.) We also mention the growing maturity of static analysis tools, that can greatly simplify both development and verification activities. By analyzing code, often while it is being written, these tools, in fact, use heuristics and rules to point out suspicious use of code constructs and thus reduce several possible sources of errors.

Versioning systems. Versioning systems are used to manage changes to software in a controlled way. Once file based (e.g., RCS and SCCS), more modern versioning systems allow for the coherent management of whole sets of files (e.g., SVN, CVS, IBM Clearcase). The underlying machinery is often based on a central repository, which stores changes history, system snapshots, and symbolic names assigned to snapshots. Commands to branch (start independent development) and merge might help simplify development in certain environments.

The new trend in the area is represented by distributed versioning systems. When using distributed versioning systems, there is no central repository and every snapshot of a software system is a repository. Changes propagate in a peer-to-peer fashion. Some popular distributed versioning systems include Mercurial, Git, and Darcs. The advantages are improved efficiency and better support for projects with large or geographically distributed teams.

Wikipedia has a page with some good starting points on the matter (Various Authors, 2009b).

Build and Continuous Integration. These tools simplify the management of complex software systems using external components and libraries, possibly evolving at different speeds. Main tasks include the management of versions and dependencies from external libraries, automated builds, and periodic integrity checks.

4.12.2 Supporting the Testing Workflow

Tools in this area can be broadly divided into three groups:

- Tools that help maintain specification of tests
- Tools to support the automatic generation and execution of tests
- Tools to support the evaluation of tests

The first area includes tools to simplify the management of the test specifications. Often available as web-based applications, these tools allow us to specify test cases, trace them back to requirements, store the results of test campaigns, and associate them to specific configurations of the system being tested. Integration with bug-tracking tools is typically provided by commercial tools. We mention Testlink, a freely available tool.

The second area includes tools, based on model checkers and static analyzers, that allow the automatic generation of test cases for software components. In this area we also mention scripting-based environments for the automation of tests. These tools help automate certain tedious tasks related to test execution by providing specification languages and execution environments to automate tests. However, it must be pointed out that the costs related to maintaining the scripts in sync with an evolving implementation might make their use inefficient. Abbot, a freely available tool for automation of GUI testing, and Various Authors (2009a) are two good starting points.

The third area, finally, includes tools to verify the completeness of tests. This is achieved by looking at code coverage, that is, by measuring how much and in which way source code has been exercised by each test case. See Chapter 5 and the discussion on DO-178B in Chapter 6 for more details.

4.12.3 Supporting the Safety Analysis Workflow

We categorize tools in this area into three different groups:

- Tools for the management of safety-related information
- Tools for the quantitative assessment of risks and hazards
- Tools for the qualitative assessment of risks and hazards

In the first area we include tools, such as FaultTree+, that provide a semantically meaningful way of drawing and managing fault trees. These tools thus assist the safety engineer in the development and management of information related to safety assessment.

The second area includes tools for the quantitative evaluation of hazards and risks. Based on probability theory and on techniques such as Monte Carlo simulation and Markov analysis, these tools support safety analysts in assigning the probability of failure to hazards and events.

The third area includes tools such as NuSMV-SA that, based on a qualitative analysis of the architecture and of the hazards, allow for the automatic generation of fault trees. See Appendix B for more details.

Tools in these three different areas can be integrated to allow for a seamless approach to the management of safety-related information. This is shown in Figure 4.18 where we highlight the information flow and an usage scenario. In particular,

- *Qualitative tools* are used to automatically generate, for instance, fault trees for a particular component (upper part of the figure).
- Results are produced and analyzed by the safety engineer using *management tools* (central part of the figure); the tool is also used to enrich the generated fault trees with quantitative information.
- *Quantitative tools* are then used to compute the probability of failures of the component.

4.12.4 Supporting the Project Management Workflow

We conclude with a brief excursion on the tools to support the project management workflow. In this area we mention Configuration Management and Issue

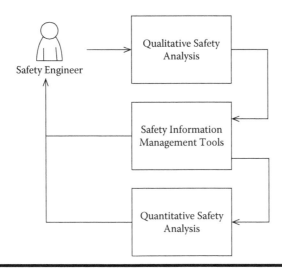

Figure 4.18   Safety analysis tools integration.

Management/Bug Tracking Tools. We leave the discussion and presentation of the tools and techniques for project management, such as Gantt chart drawing, EVA analysis, risk assessment tools, to books specifically dedicated to the subject (see, e.g., Burke (2006), Office of Government Commerce (2005), and Project Management Institute (2004)).

Configuration Management Tools. This kind of activity is supported by tools that ensure a proper versioning and management of a coherent set of information about a system. The concept extends that of a versioning system by implementing detailed workflows to support change and by allowing for a more rigorous definition of what constitutes a configuration.

For software development, tools such as IBM Clear Quest support and integrate configuration management in development environments.

Issue and Bug-tracking Management. It includes tools to keep track of issues raised and found during the various phases of development. The tools, typically web based, allow us to associate various attributes to bugs and link them to specific configurations and versions of the system under development. For software development, tools such as Bugzilla and Mantis are two freely available and widely used choices.

4.13 Improving the Process: Maturity Models

Various models exist to measure the maturity of organizations in developing systems (e.g., CMMI (CMMI Product Team, 2009); ISO-IEC 15504 (ISO/IEC 15504, 1998); Bootstrap (Kuvaja et al., 1994); and Trillium (Bell Canada, 1994)). Among them, the CMMI is of particular interest both for its wide adoption and for the fact that extensions have been defined for the development of safety-critical applications (Defence Materiel Organisation, Australian Department of Defence, 2007). CMMI (Capability Maturity Model Integration) is, in fact, a family of models meant to cover different aspects of production and service delivery. The model was first defined by the Software Engineering Institute (1995).

The CMMI distinguishes five different levels, corresponding to increasing capabilities[7]. Each level relies on the lower levels, so that in order to achieve level 4, for instance, all the practices defined for level 3 must be in place:

1. **Initial,** when there are no specific procedures in place for the delivery of products and services.
2. **Managed,** when a basic infrastructure is in place to support the production of systems. The discipline reflected by this capability level helps ensure that existing practices are retained during times of stress.
3. **Defined,** when development and production processes are defined and tailored from the organization's set of standard processes according to the organization's tailoring guidelines. Notice that this is the first level of interest for the development of safety-critical applications, because achieving certification of a product is unlikely if production is performed at level 1 or 2.
4. **Measured,** when a quantitative system is in place to statistically control and evaluate performance and quality.
5. **Optimized,** when the statistical data can be effectively used to improve the production process.

The capability level is assessed by measuring the maturity of the organization in different *Process Areas* (PAs). PAs identify critical areas and practices for the development of systems. The CMMI defines nineteen process areas, although not all of them apply to all maturity levels. They are summarized in Table 4.9 for generic development and in Table 4.10 for safety analysis.

---

[7] They are called "capability" and "maturity" levels. We will just call them "maturity" levels. Moreover, the continuous path also introduces a level zero, standing for not defined.

Table 4.9   CMMI Process Areas

CMMI PA Category	Safety Process Area
Engineering	Requirements Development
	Requirements Management
	Technical Solution
	Product Integration
	Verification
	Validation
Project Management	Project Planning
	Project Monitoring and Control
	Supplier Agreement Management
	Integrated Project Management + IPPD
	Risk Management
	Quantitative Project Management
Process Management	Organizational Process Focus
	Organizational Process Definition + IPPD
	Organizational Training
	Organizational Process Performance
	Organizational Innovation and Deployment
Support Processes	Configuration Management
	Process and Product Quality Assurance
	Measurement and Analysis
	Decision Analysis and Resolution
	Causal Analysis and Resolution

Table 4.10 CMMI+SAFE Process Areas

CMMI PA Category	Safety Process Area	Specific Goals
Project Management	Safety Management	SG1. Develop safety plans
		SG2. Monitor safety incidents
		SG3. Manage safety-related suppliers
Safety Engineering	Safety Engineering	SG1. Identify hazards, accidents, and sources of hazards
		SG2. Analyze hazards and perform risk assessments
		SG3. Define and maintain safety requirements
		SG4. Design for safety
		SG5. Support safety acceptance

References

AIA, GAMA, and FAA Aircraft Certification Service (2004). The FAA and Industry Guide to Product Certification. Available at http://www.faa.gov/aircraft/aircert/design−approvals/media/CPI−guide−II.pdf. Last retrieved on November 15, 2009.

AIRBUS (Last retrieved on November 15, 2009). A380 Family. Available at http://www.airbus.com/en/aircraftfamilies/a380/.

Albrecht, A.J. (1979). Measuring application development productivity. In *Proc. IBM Application Development Symposium*, pp. 83–92. Indianapolis, IN: IBM Press.

Anschuetz, L. (1998). Managing geographically distributed teams. In *A Contemporary Renaissance: Changing the Way We Communicate, Proc. International Professional Communication Conference (IPCC 98)*. Washington, D.C.: IEEE Computer Society.

BEA (2002). Accident on 25 July 2000 at La Patte d'Oie in Gonesse (95) to the Concorde Registered F-BTSC operated by Air France (Main Report). Available at http://www.bea-fr.org/docspa/2000/f-sc000725a/pdf/f-sc000725a.pdf. Last retrieved on November 15, 2009.

Bell Canada (1994). The Trillium Model. Available at http://www2.umassd.edu/swpi/BellCanada/trillium-html/trillium.html. Last retrieved on November 15, 2009.

Boehm, B.W. (1981). *Software Engineering Economics (Prentice-Hall Advances in Computing Science & Technology Series)*. Upper Saddle River, NJ: Prentice Hall.

Boehm, B.W. (1988). A spiral model of software development and enhancement. *IEEE Computer* 21(5), 61–72.

Boehm, B.W., E. Horowitz, R. Madachy, D. Reifer, B.K. Clark, B. Steece, W.A. Brown, S. Chulani, and C. Abts (2000). *Software Cost Estimation with Cocomo II (with CD-ROM)*. Upper Saddle River, NJ: Prentice Hall.

Booch, G., J. Rumbaugh, and I. Jacobson (2005). *Unified Modeling Language User Guide* (2nd ed.). Object Technology Series. Reading, MA: Addison-Wesley.

Burke, R. (2006). *Project Management, Planning and Control Techniques (4th ed.)*. New York: Wiley.

Cai, X., M.R. Lyu, and M.A. Vouk (2005). An experimental evaluation on reliability features of n-version programming. In *Proc. 16th IEEE International Symposium on Software Reliability Engineering (ISSRE'05)*, pp. 161–170. Washington, D.C.: IEEE Computer Society.

Clarke, E.M., O. Grumberg, and D.E. Long (1992). Model checking and abstraction. In *POPL '92: Proc. 19th ACM SIGPLAN-SIGACT Symposium on Principles of Programming Languages*, New York, pp. 343–354. ACM.

Clarke, E.M., O. Grumberg, and D.E. Long (1994). Model checking and abstraction. *ACM Transactions on Programming Languages and Systems (TOPLAS)* 16(5), 1512–1542.

CMMI Product Team (2009). Capability Maturity Model Integration. Technical Report CMU/SEI-2006-TR-008. Pittsburgh, PA: Software Engineering Institute.

COCOMO II (2000). COCOMO II Model Definition Manual. Available at http://csse.usc.edu/csse/research/COCOMOII/cocomo2000.0/CII-modelman2000.0.pdf. Last retrieved on November 15, 2009.

Conkey, C. and A. Pasztor (2009). U.S. Aviation Faces Surging Bird-Strike Threat. Available at http://online.wsj.com/article/SB124058077567352845.html. Last retrieved on November 15, 2009.

Dassault Systemes (Last retrieved on November 15, 2009). Catia - Virtual Design for Product Excellence. Available at http://www.3ds.com/products/catia/welcome/.

Defence Materiel Organisation, Australian Department of Defence (2007). +SAFE, V1.2: A Safety Extension to CMMI-DEV, V1.2. Technical Report CMU/SEI-2007-TN-006, Pittsburgh, PA: Software Engineering Institute.

Department of Defense (1993). Military Standard—System Safety Program Requirements. Technical Report MIL-STD-882C, Department of Defense.

Department of Defense (1998). Department of Defense Handbook—Work Breakdown Structure. Technical Report MIL-HDBK-881, Department of Defense.

Department of Defense (2000). Department of Defence, Standard Practice for System Safety. Technical Report MIL-STD-882D, Department of Defense.

Dörner, D. (1997). *The Logic of Failure: Recognizing and Avoiding Error in Complex Situations*. New York: Perseus Books.

Dvorak, D.L., Editor (2009). Nasa study on flight software complexity. Available at http://oceexternal.nasa.gov/OCE-LIB/pdf/1021608main-FSWC-Final-Report.pdf. Last retrieved on November 15, 2009.

Embry-Riddle Aeronautical University (Last retrieved on November 15, 2009). FAA National Wildlife Strike Database. Available at http://wildlife.pr.erau.edu/database/mapping-us-select.php.

European Space Agency (2008). Space Project Management: Project Planning and Implementation. Technical Report ECSS-M-ST-10C. Noordwijk, The Netherlands: European Cooperation for Space Standardization (ECSS).

FAA (Federal Aviation Administration) (2000). FAA System Safety Handbook. Available at http://www.faa.gov/library/manuals/aviation/risk-management/ss-handbook/. Last retrieved on November 15, 2009.

FAA (Federal Aviation Administration) (2007). Type Certification. Order 8110.4C, U.S. Department of Transportation.

FAA (Federal Aviation Administration) (Last retrieved on November 15, 2009). Airport Wildlife Hazard Mitigation Home Page. Available at `http://wildlife-mitigation.tc.faa.gov/public_html/index.html`.

Fowler, M. and K. Scott (2000). *UML Distilled (2nd ed.): A Brief Guide to the Standard Object Modeling Language.* Reading, MA: Addison-Wesley.

Giunchiglia, F., A. Villafiorita, and T. Walsh (1999). Theories of abstraction. *AI Communications* 10(3-4), 167–176.

Giunchiglia, F. and T. Walsh (1992). A theory of abstraction. *Artificial Intelligence* 57(2-3), 323–389.

Hinds, P.J. and M. Mortensen (2005). Understanding conflict in geographically distributed teams: The moderating effects of shared identity, shared context, and spontaneous communication. *Organization Science*, 16(3), 290–307.

INCOSE (Last retrieved on November 15, 2009). Requirements Management Tools Survey. Available at `http://www.incose.org/ProductsPubs/Products/rmsurvey.aspx`.

ISO/IEC 15504 (1998). ISO/IEC 15504 : Information Technology—Software Process Assessment—part 7: Guide for Use in Process Improvement.

Joint Software System Safety Committee (1999). Software System Safety Handbook—A Technical & Managerial Team Approach. Available at `www.system-safety.org/Documents/Software_System_Safety_Handbook.pdf`. Last retrieved on November 15, 2009.

Jones, C. (2007). *Estimating Software Costs.* New York: McGraw-Hill.

Knight, J.C. and N.G. Leveson (1986). An experimental evaluation of the assumption of independence in multiversion programming. *IEEE Transactions on Software Engineering,* 12(1), 96–109.

Kruchten, P. (2000). *The Rational Unified Process: An Introduction* (2nd ed.). Reading, MA: Addison-Wesley.

Kuvaja, P., J. Simila, L. Krzanik, A. Bicego, G. Koch, and S. Saukonen (1994). *Software Process Assessment and Improvement: the BOOTSTRAP approach.* New York: Blackwell.

Lann, G.L. (1997). An analysis of the Ariane 5 flight 501 failure—a system engineering perspective. In *Proc. IEEE International Conference and Workshop on Engineering of Computer-Based Systems,* pp. 339–346. Washington, D.C.: IEEE Computer Society.

Meyer, B. (1992). Applying "design by contract." *Computer* 25, 40–51.

Meyer, B. (2009). *Touch of Class: Learning to Program Well with Objects and Contracts.* Berlin: Springer.

Miller, A. (2008). Distributed Agile Development at Microsoft Patterns & Practices. Available at `http://download.microsoft.com/download/4/4/a/44a2cebd-63fb-4379-898d-9cf24822c6cc/distributed_agile_development_at_microsoft_patterns_and_practices.pdf`. Last retrieved on November 15, 2009.

MRY/Reuters (2006). A380 hit by new production problems. *Spiegel Online International.* Last retrieved on November 15, 2009.

NASA (1999). Mars Climate Orbiter Mishap Investigation Board Phase I Report. Available at `ftp://ftp.hq.nasa.gov/pub/pao/reports/1999/MCO_report.pdf`. Last retrieved on November 15, 2009.

NASA (2004). *NASA Software Safety Guidebook.* Number NASA-GB-8719.13 in NASA Technical Standard. National Aeronautics and Space Administration.

NASA (2007). NASA Systems Engineering Handbook. Technical Report NASA/SP-2007-6105, Rev1, NASA.

NEi Software Inc. (Last retrieved on November 15, 2009). NEi Software Automotive Case Study. Available at `http://www.nenastran.com/newnoran/chPDF/CASE-Chassis-Design.pdf`.

O'Connor, P.D. (2003). *Practical Reliability Engineering (4th ed.)*. New York: Wiley.

Office of Government Commerce (2005). *Managing Successful Projects with PRINCE2 (5th revised ed.)*. Washington, DC: Stationery Office Books.

Perrow, C. (1984). *Normal Accidents: Living with High-Risk Technologies*. Basic Books. Updated by Princeton, NJ: Princeton University Press, 1999.

Pilone, D. and N. Pitman (2005). *UML 2.0 in a Nutshell (In a Nutshell (O'Reilly))*. Sebastopol, CA: O'Reilly Media, Inc.

Project Management Institute (2004). *A Guide to the Project Management Body of Knowledge (PMBOK®) Guide) (3rd ed.)*. Newtown Square, PA: Project Management Institute.

Radu, G. and D. Mladin (2003). Common cause failure data collection and analysis for safety-related components of TRIGA SSR-14MW Pitesti, Romania. In *Proc. International Conference Nuclear Energy for New Europe 2003*.

Report by the Inquiry Board (1996). Ariane 5 Flight 501 Failure. Paris, France: European Space Agency.

Royce, W. (1998). *Software Project Management: A Unified Framework*. Reading, MA: Addison-Wesley.

Royce, W.W. (1970). Managing the development of large software systems. In *Proc. Western Electronic Show and Convention (WESCON 1970)*, pp. 1–9. IEEE Computer Society. Reprinted in *Proc. 9th International Conference on Software Engineering*, Toronto, Ontario, Canada: ACM Press, 1989, pp. 328–338.

SAE (1996a). Certification Considerations for Highly-Integrated or Complex Aircraft Systems. Technical Report ARP4754. Warrendale, PA: Society of Automotive Engineers.

SAE (1996b). Guidelines and Methods for Conducting the Safety Assessment Process on Civil Airborne Systems and Equipment. Technical Report ARP4761. Warrendale, PA: Society of Automotive Engineers.

Software Engineering Institute (1995). *The Capability Maturity Model: Guidelines for Improving the Software Process*. Reading, MA: Addison-Wesley.

Sommerville, I. (2007). *Software Engineering (8th ed.)*. Reading, MA: Addison-Wesley.

University of Southern California (1994). *USC COCOMO Reference Manual*. Los Angeles, CA: University of Southern California.

University of Southern California (2000). USC COCOMO II 2000 Software Reference Manual. Available at `http://csse.usc.edu/csse/research/COCOMOII/cocomo2000.0/CII-manual2000.0.pdf`. Last retrieved on November 15, 2009.

University of Southern California (Last retrieved on November 15, 2009). Center for Systems and Software Engineering of the University of Southern California. Available at `http://csse.usc.edu/csse/index.html`.

Various Authors (Last retrieved on November 15, 2009a). List of GUI Testing Tools. Available at `http://en.wikipedia.org/wiki/List-of-GUI-testing-tools`.

Various Authors (Last retrieved on November 15, 2009b). Revision Control. Available at `http://en.wikipedia.org/wiki/Revision-control`.

Chapter 5

Formal Methods for Safety Assessment

5.1 Introduction

The term **formal methods** refers to the use of formalisms and techniques, based on mathematics, for the specification, design, and analysis of computer hardware and software. Formal methods are based on the use of mathematical logic to formalize the reasoning process. Mathematical logic includes, for instance, the use of propositional or predicate calculus, or more complex higher-order logics. Formal methods must be distinguished from **formalized methods** (Rushby, 1993). Formalized methods include several notations, such as different sorts of diagrams, or pseudocode, that have a semblance of formality but typically do not rely on formal underlying semantics.

Traditionally, formal methods are based on the use of discrete mathematics (Storey, 1996; Rushby, 1993). Instead of computing values for physical quantities that are continuous in nature, formal methods prove theorems about their behavior. For instance, formal methods can show that the temperature T satisfies the condition that $70 \leq T \leq 180$, instead of computing the value of the temperature according to some physical laws described in continuous mathematics. The use of discrete mathematics is particularly suitable for reasoning about digital systems that are discrete in nature; that is, their behavior can be described as a succession of discrete state changes. However, more recent trends include the application of formal methods for modeling and verification of analog systems (or combinations of analog and digital systems). An example of this trend is given by recent works on hybrid systems (compare Section 5.10).

The areas of application of formal methods are manifold. First of all, they include the use of formal languages, that is, languages endowed with formally defined syntax and semantics, for system modeling. Second, formal methods include the use of formal techniques to analyze system models. For instance, formal methods can be used to prove that a system model satisfies a set of requirements. Formal methods for system analysis are based on mathematical techniques that, given a set of definitions and assumptions, are able to produce a proof for a given argument, based on a rigorous deduction process. As an example, in theorem proving (see Section 5.7.2), a property can be shown to be a logical consequence of a set of axioms, if it can be formally derived from the axioms with a set of deduction steps, that are instances of the set of inference rules that are allowed in the language. Furthermore, every formal proof can be checked for correctness using a mechanical process. Alternative formal techniques include the use of methods that are able to check the correctness of a given property via symbolic examination of the entire state space of a given system, hence proving that the property is true for all possible inputs (compare Section 5.7.3). Finally, formal methods can be used to support and guide the system development process, from the early phases of requirement elicitation to final system implementation. Application of formal methods in the development life cycle includes, for instance, verification that different models of the same system, at different levels of detail (e.g., an architectural model and the corresponding detailed design model), enjoy some correctness relationship; for instance, the design model is a refinement of the architectural one. The application of formal methods in the development process is discussed in Section 5.3.

In this chapter we look at the application of formal methods for the verification and validation of complex systems, which may include both hardware and software components. We first discuss the advantages and limitations related to the use of formal methods, and present a brief historical perspective of their use. Finally, we look in more detail at some of the most well-known formalisms for the specification and analysis of complex systems. The use of formal methods is not limited to verification of correctness. They can also be used to address safety-related activities. In Section 5.8 we discuss the application of formal methods to formal safety analysis.

5.2 Advantages of Formal Methods

The use of formal methods is motivated by the growing complexity of computer-based systems. Computer-based systems are becoming more and more complex, both in the way they interact with the environment and in the type of functionalities they provide. This is true for systems that are part of our everyday life (e.g., cellphones) as well as for highly critical systems (e.g., fly-by-wire control systems or shutdown systems for nuclear power plants). Formal methods provide rigorous methodologies and techniques that can aid the specification, design, and analysis of these systems. The goal of using formal methods is to increase confidence in the correctness and safety-related features of critical systems, reduce time-to-market, and reduce costs related to system development, maintenance, and error correction.

The first motivation for using formal methods is as a methodology to improve system specification and modeling. Formal methods can remove ambiguities in the specification of both system requirements and system models. As a difference from traditional specification methods, which very often rely on natural language, formal methods are based on formal languages that must obey very precise rules. The resulting specifications have very precise semantics and therefore can be interpreted and checked for correctness unambiguously. Moreover, formal specifications can help in the sharing of information between different actors contributing to system development (e.g., design engineers and safety engineers). Informal methodologies rely on engineering judgment and expertise, whereas formal methods eliminate the risk of misunderstandings or wrong interpretations, that often arise when more traditional methodologies are used. Specifications produced using formal methods are well documented and ease product maintainability, reusability, and change.

However, the area in which the use of formal methods may have a crucial advantage over traditional techniques is for system verification and validation. In this context, the term **verification** refers to checking whether a given system correctly implements the stated requirements, whereas the term **validation** refers to checking whether the stated requirements are the intended ones, that is, the system conforms to user expectations. Traditional techniques for system verification include testing and simulation (see Section 5.7.1). A drawback of testing is that it is not exhaustive. Testing provides conclusive results only for the test patterns that have been examined, and does not guarantee system correctness for the untested ones. This is an even more serious concern when testing is applied to digital systems that admit discontinuities (i.e., a discrete state change may result in an abrupt change of the value of a given variable). In such cases, the fact that the system behavior is correct for an input pattern does not allow us to infer that it will still be correct for a different pattern that is "close" (whatever this means) to the previous one. Formal methods have the potential to deal with these issues, in that they enable an exhaustive and conclusive analysis of the system at hand. Very often, formal methods provide tool support, and in some cases the verification process can be completely automatized. In the context of validation, formal methods provide techniques to check the consistency and completeness of sets of requirements. Additionally, they provide prototyping and animation functionalities that can be used to to check the adequacy of a specification with respect to user expectations. Finally, formal methods can be used to assist the system development life cycle. The advantages of formal methods in system development are discussed in more detail in the next section.

5.3 Formal Methods in the Development Process

Formal methods can be used in a number of ways and at different levels of formality within the development life cycle. In principle, they can pervade every single phase of the development. Formal methods can be used for requirements capture and system specification at different levels of detail, from architectural design to detailed design and the final implementation level. An advantage of using formal methods (as opposed to, for example, pseudocode) for the system design phase is that formal languages can be used to describe the system at hand in a sufficiently generic manner, so that it is possible to delay design choices related to the final realization of the system. For instance, it is possible to delay the choice of whether a particular functionality will be realized in hardware or in software. Formal methods can also be used to formally relate the outcomes of the different phases—for instance, to prove the equivalence of requirements and specifications at different levels of abstraction—and resolve errors in specification refinements. Finally, formal methods can be used to support incremental system development. There are different levels of rigor in the application of formal methods. Rushby (1993) classifies their use into four different levels, as follows:

Level 0 *No use of formal methods.* This level corresponds to the standard practice. Specifications are written in natural language, or make limited use of formalized notations, such as diagrams and pseudocode. Verification and validation are based on manual review and inspection, simulation, and testing.

Level 1 *Use of concepts and notation from discrete mathematics.* This level makes use of concepts and notation from discrete mathematics and mathematical logic, such as set theory, functions, and relations, to formalize requirements and specifications. Proofs are typically carried out informally, similar to what is typically done in mathematics textbooks. Advantages of this approach include a more precise and compact presentation of specifications and requirements, and the possibility of using mathematical reasoning to relate specifications at different levels. The formal approach is typically performed in parallel with, but does not replace traditional development methodologies.

Level 2 *Use of formalized specification languages with some mechanized support tools.* With respect to the previous level, here formal notations and techniques with a fixed concrete syntax and semantics are used. The languages typically have at least limited tool support. For instance, formal languages may be equipped with syntax and type checkers. Proofs are still performed manually. However, some languages may provide tool support for formal deduction, based on inference rules. The term *rigorous*, as opposed to *formal*, is often used to denote proofs at this level. Level 2 methodologies improve the benefits of level 1 by providing languages that address software engineering issues and mechanized checking tools to detect some types of faults in the specifications. Finally, they may enable prototyping or animation of formal specifications.

Level 3 *Use of fully formal specification languages with comprehensive support environments, including mechanized theorem proving or proof checking.* The highest level of rigor of the application of formal methods consists of using fully formal specification languages that are supported by tools that enable automatic generation of proofs. Proofs are fully formalized and can be produced automatically (either with human guidance, such as in interactive theorem proving, or without human guidance, such as in automated theorem proving or in model checking). The advantages of level 3 methods lie in the fact that mechanical proofs eliminate the possibility of wrong reasoning due to human error (although they can evidently be affected by bugs in the tools themselves), and increase the ability of detecting specification faults. On the negative side, fully formal notations and methods may sometimes be difficult to use, and they may have a significant impact in terms of development costs.

Formal methods with a higher level of rigor are not necessarily better than lower-level ones. The choice of which level of rigor in the application of formal methods should be used depends on a number of factors that are peculiar to the specific project at hand (e.g., the safety integrity level of the application being developed). For instance, lower-level methods might be employed for the development of systems where safety integrity is not the main concern. Using a limited amount of formal techniques in such projects can still help in producing well-designed and well-documented systems, thereby improving maintainability. Formal methods at the highest level of rigor may be reserved only for the development of highly critical systems, where the additional costs incurred when using fully formal methods are justified by the need to have a greater degree of assurance in the safety-related features of the system. Typically, the final decision about which level of rigor should be used represents a trade-off between the benefits that are expected and the additional costs and time that may be required by their use.

Within a project, it is also possible to vary the extent of application of formal methods. First of all, formal methods need not be applied everywhere. It is possible to apply them only at selected stages of the development life cycle; that is, use formal methods to automate some phases of the development and resort to traditional techniques for the remaining stages. A possible objection is that if formal methods are not applied pervasively in the life cycle, then the overall argument about correctness of the design must still rely on standard techniques, such as review and inspection. On the other hand, effectiveness, feasibility, and cost considerations might suggest limiting the application of formal methods only to specific stages. In the literature, two alternative approaches have been advocated, each of them with different implications, advantages, and drawbacks: apply formal methods *early* in the development life cycle, or apply them *late*. Proponents of use at the later stages of development argue that it is the final implementation-level design that must be verified. If the final product, be it a piece of software code or a gate-level design or a combination of both, has not been verified, then the verification process is useless. On the other hand, there are several reasons for applying formal methods at early stages. First, verification that the final implementation conforms to its specification is pointless if the specification itself turns out to be flawed. Second, it is well known that bugs introduced during the early stages of the design are the most dangerous, and fixing them is very expensive when they are discovered late in the development. Finally, proponents of the use of formal methods at the early stages argue that formal methods are more fruitfully applied to phases where traditional techniques are weaker. Typically, simulation and testing can be very effective in discovering bugs at code or gate level. However, testing often comes too late in the design, and the early stages of development must rely only on manual review and inspection, which is carried out on informal specifications. Formal methods, on the other hand, have the potential to deal in a better way with verification and requirements validation at early stages in the design. Finally, using formal methods for implementation-level specifications is typically more costly, and sometimes infeasible, simply because of the dimension and level of detail of the specification itself. Architectural-level specifications typically specify system behavior in purely functional terms, hence falling into the realm of discrete mathematics and logic, whereas implementation-level specifications may require reasoning in terms of physical quantities using continuous mathematics.

In addition to applying formal methods only to selected stages of the design, it is possible to limit their application only to selected subsystems or components, instead of the complete system. For instance, it is possible to restrict the use of formal techniques (in particular, of those with the highest level of rigor) only to components that have the highest levels of criticality. Components that are not safety critical, on the other hand, can be verified with more traditional methods.

A further possibility is to limit the application of formal methods only to selected verification tasks, and to system properties other than full functionality. In principle, verification involves proving that the complete design is correct and complete with respect to its specification. Hence, proofs of correctness are carried out to verify the functional behavior of each component and of the overall system. However, in general it may be the case that not all functional properties have the same degree of importance. In safety-critical systems, for instance, it may sometimes be more important to show that a component does *not* show a given behavior, or that it does not fail in a catastrophic and unrecoverable manner, instead of showing that it can guarantee availability of the function it is designed to perform. A possible development choice is to restrict the application of formal methods, in the case of safety-critical systems, to only the verification of selected safety properties, and perhaps to the generation or artifacts that are typical of safety analysis, such as fault trees and FMEA tables. The application of formal methods for safety assessment is discussed in Section 5.8.

We conclude this section by noting that formal methods can be used to support *certification* of safety-critical systems. Certification is typically a necessary step that must be achieved before a safety-critical system can be deployed and used. In many industrial sectors (e.g., in civil avionics or in nuclear power generation), certification is always compulsory. Even when not compulsory, certification can be sought in order to enlarge the potential market for a given product. Certification requires convincing a regulatory authority of the appropriateness and safety level of a given product. Typically, the set of requirements and guidelines that the product must satisfy (including, e.g., compliance with specific development practices and a specific safety life-cycle model) are set forth by regulatory bodies in a so-called *standard.* Some standards are specific to a particular sector, whereas others are more generic. In the documents describing some of these standards, formal methods are recommended, or at least admitted as an alternative with respect to more traditional techniques, to carry out (parts of) the system development. The role of formal methods for certification is discussed in detail in Chapter 6.

Before entering into the description of specific formalisms and techniques, in the next section we discuss some problems and limitations that are related to the use of formal methods.

5.4 Problems and Limitations

Formal methods are not exempt from problems and limitations. As a matter of fact, over the years there has been strong debate between supporters and detractors of formal methods, a debate that is not yet over. Objections set forth against the use of formal methods, at least over the past decades, included the limited tool availability and the insufficient level of maturity of most techniques. However, the situation has currently improved significantly compared to years ago (Hall, 1990; Bowen and Hinchey, 1995). Nevertheless, there are some limitations and restrictions in the use of formal methods that cannot be simply ignored. In this section we discuss some of these issues.

A serious argument against formal methods is that their application can be expensive; this may increase project costs and slow down the development process. Timing constraints can sometimes hinder the application of formal methods, especially the ones with the highest level of rigor. As already mentioned in Section 5.3, the use of fully formal methods is typically restricted to the most critical projects, and is justifiable only for a limited number of safety-critical applications. However, in such applications, the additional burden of using formal methods is typically compensated by the higher degree of assurance in the safety and correctness of the system being developed. Furthermore, in some applications the costs incurred by using formal methods from the early stages of system design are justified by the significantly higher costs for correcting undetected bugs at later stages.

Detractors of formal methods often argue that formal methods are difficult to use. The expressiveness of some formal languages is balanced by the complexity of the specifications that can be produced. Using formal methods and mathematical techniques requires considerable expertise, and engineers do not always have the necessary level of training to master their use. Expertise is required not only to write good specifications in formal languages, but also to use tools for formal verification. Verification using formal tools is often far from being automatic. Sometimes, logically equivalent specifications can result in significantly different tool performances when fed into the same verification tool. Hence, expertise may be required in order to produce the specification that is more amenable to automatic verification, although it is often possible to provide some general guidelines for the specifier. The solution to this state of things is simply to have education and training in formal methods as a necessary background for design and safety engineers. Although this process is far from being completed, especially in some industrial sectors, this is the direction that industry is currently going.

Another objection related to the use of formal methods is their limited applicability. To achieve maximum benefit, formal methods should be applied at each stage of the development process, which is not always possible due to limitations of current techniques or other development constraints. We already mentioned this issue in Section 5.3, where we outlined some possible solutions. If necessary, formal methods can be limited to specific phases or components, or to verification of selected properties. Although this is clearly a limitation of their applicability, using formal methods can still provide greater confidence in the results of verification. Furthermore, a peculiarity of formal methods is that, even when applied only for requirements and system specification, they help engineers acquire a deeper understanding of the system being modeled. This is an invaluable advantage, regardless of whether or not the modeled system will be eventually verified and validated using formal techniques. In fact, the use of so-called **lightweight formal methods** has been sometimes advocated. Lightweight formal methods consist of the limited application of formal methods whose goal is to produce partial specifications of the systems being modeled, and to perform focused application of verification techniques to tests for errors and bugs in the specifications. Lightweight formal methods lack the expressive power and coverage that fully formal methods have. However, what matters in their application is the additional insight that is gained during the modeling and validation processes.

An issue that is again related to the development process is the fact that using a single formalism for every stage of the process may be problematic. There are formal languages that are more suitable for describing the architectural design of a given system, and yet other formal languages that are more suitable to describe its implementation-level design. Similarly, there may be languages that are good at describing software designs, and others that are better at describing hardware designs. Sometimes, constraints may be posed by the fact that some languages are the *de facto* standard in industry for some specific applications and cannot easily be replaced. Using different languages for different phases of the development is clearly an option, and in fact this has been done in several projects where formal methods have been used. On the other hand, such use of formal languages may pose problems when the results of verification must be put together to produce the overall safety argument. This is especially true when formal methods at the highest level of rigor are being used.

Formal verification, as already mentioned, is far from being automatic. Furthermore, it is not always effective or conclusive. This may be due to the dimension of the verification problem at hand, or to the inability of the formal verifier to decide about some classes of verification problems. The solution of large problems may require computing resources (in terms of computation time, or memory requirements) that are not available or simply unacceptable. An example of this is the so-called *state explosion problem* that limits the applicability of techniques that carry out automatic verification by exhaustively exploring the state space of a given system, either explicitly or using symbolic techniques (see Section 5.7.3). Other formal techniques may be inconclusive, simply because they are not powerful enough to deal with some classes of problems. Apart from technological advances, possible solutions to this problem include adapting existing technologies to the specific problem at hand, or simplify the verification problem by adapting the specification (e.g., using abstraction techniques to abstract away unnecessary details).

Further arguments against formal methods often include the fact that formal specifications could be mistaken. A correctness result stating that a given property holds for some formal specification is clearly meaningful only to the extent that the specification correctly captures the designer's intent. This problem can be alleviated when using specification languages that are executable. Such languages typically provide the designer with emulation and animation capabilities for exploring the behavior of the specification. An additional problem is related to the fact that a specification is always necessarily an abstraction of the real system it models. Verification could miss finding a bug for the sole reason that the specification might not possess the level of detail necessary to uncover the bug. Hence, verification results may not be always accurate. However, this is not a distinguishing feature of formal specifications with respect to informal ones. The crucial difference is that formal methods, as opposed to informal ones, force the specifier to state explicitly all the assumptions and constraints that must hold for the results of the verification process to be trustable.

Some detractors of formal methods claim that the results of verification (e.g., a proof generated by a theorem prover) may not always be the kind of artifacts that are accepted by human reviewers as evidence for correctness. Nevertheless, an invaluable advantage of formal verifiers is that they require the user to formally state all the preconditions and assumptions that sustain a given proof, and, as a consequence, they give useful insight into the reasons for the correctness or incorrectness of a given argument. Similarly, model checkers (see Section 5.7.3) can generate counterexamples whenever a given argument is disproved, thus helping the designer pinpoint the problem with the specification being verified. In addition, automated tools are very good at carrying out tedious and mechanical verification steps that can be subject to errors when performed manually. Overall, formal verification should not be seen as something that replaces human review, but rather as a way to enhance it.

A further argument against formal verification is the possibility that the formal verifier may produce wrong results, for instance, due to a software bug affecting it or due to a bug in the compiler used for object code generation. Use of certified and validated compilers can mitigate, but not fully resolve this issue. A validated compiler is tested and checked for compliance with the standards, but, as any other piece of software, can still contain faults. Finally, and more importantly, it should be noted that the goal of formal verification is not to deprive human users of their responsibility, but rather to enhance their ability to carry out intricate and error-prone verification tasks.

5.5 History of Formal Methods

The use of formal methods for system design and verification dates back to the 1970s. At that time, formal methods played a significant role in the field of *security*, rather than *safety-critical* systems (Wing, 1998). The National Security Agency in the United States (either directly or via the National Computer Security Center) and its Canadian and United Kingdom counterparts were the main sources of funding in formal methods at the time. Development and verification of secure systems and applications proved a very challenging area. At first, formal methods were used to prove security properties of operating systems. In particular, the research focused on operating system kernels, properties such as data integrity, and properties related to access rights. Formal methods were then applied to reason about security protocols and prove related security issues (e.g., secrecy). Formal methods helped identify previously unknown flaws in several security protocols. The main achievements of formal methods in this field are mainly due to the development of verification tools based on theorem proving (compare Section 5.7.2).

During the early 1980s and then the 1990s, the use of formal methods spread to other areas, such as real-time control and, more generally, safety-critical systems. It is worth mentioning that the United States, Canada, and the United Kingdom, through their governmental agencies previously mentioned, sponsored two international workshops *FM'89* (Craigen and Summerskill, 1989) and *FM'91*. These two workshops stimulated debate on the use of formal methods for both specification and verification of security-critical and safety-critical computer systems. During the 1990s, significant advances and achievements were reached also by the use of automated verification tools based on model checking (compare Section 5.7.3). By then, formal methods had been applied for a number of applications in the area of safety-critical systems. Formal methods have been used in real industrial projects, with different extents of applications to support the development and verification of safety-critical systems. We present a collection of some of the most significant examples and achievements in Section 5.9. At present, formal methods are increasingly being applied in several industrial sectors. The use of formal methods is also recommended, or least mentioned, in different standards (compare Chapter 6). Although this process is far from being completed, it is clear that formal methods have proved to be a clear success, at least in some industrial sectors. For instance, in hardware verification the use of formal methods is consolidated. Major companies such as Intel use a plethora of formal methods techniques to prove their designs correct before the production phase, and to detect bugs as early as possible. These companies routinely hire people trained in formal methods, build their own verification tools, and also have their own research departments to carry out edge research in this field.

5.6 Formal Models and Specifications

Several formal specification languages have been used and are referred to in the literature on formal methods. Examples range from languages that use algebraic notations, to various forms of process calculi, languages based on first-order logic or predicate calculus, and model-based specification languages. In this section we review and try to classify some of the most well-known formalisms for model specification. We distinguish four broad categories: **algebraic languages**, **model-based languages**, **process algebras and calculi**, and **logic-based languages**. In practice, the classification is not always clear, and some languages may incorporates features that pertain to different classes (Barroca and McDermid, 1992). Formal languages can also be classified in terms of their semantical foundation. Some languages are based on denotational or operational semantics. Other languages are based on different forms of axiomatic semantics; for instance, they may use preconditions and postconditions that can take the form of assertions inside specifications.

Specification languages are also characterized by the level of mechanization, and the kind and number of support tools that are available for them. Examples of support tools are *parsers, typecheckers, consistency checkers,* tools for *static analysis,* and different types of verification tools, including, for example, *proof checkers* or *theorem provers.* Some specification languages are executable. An advantage of executable languages is that they can be used for animating specifications, and for simulation and prototyping. A drawback is that executable languages are typically closer to programs rather than specifications, and hence can make it more difficult to abstract from the implementation details of the system being modeled, without making premature commitments (e.g., on the system being implemented in hardware or in software).

Finally, we note the difference between **property-oriented** and **model-oriented** specification styles. In property-oriented specifications, the system being modeled is specified in terms of the properties that any implementation of the system must satisfy. For example, a data structure can be modeled by means of a set of properties that axiomatize the effect of the corresponding operations. In the model-oriented specification style, a mathematical model corresponding to a particular implementation of the system is provided. For example, a particular implementation of a data structure and its operations may be given. Typically, model-oriented specifications are closer to the final implementation, and hence tend to be used in the later stages of system design. Formal verification techniques can be used to demonstrate compliance between different sorts of specifications, for instance to show that a model-oriented specification satisfies the axioms of a property-oriented specification of the same system.

5.6.1 Algebraic Specification Languages

A distinguishing feature of algebraic specification languages is to characterize the objects to be specified in terms of the algebraic relationships between them. Very often, the state of the objects is abstracted away, and the specification focuses on the algebraic relationships that define the behavior of the operations that can be performed on the objects. Programs are seen as many-sorted algebras consisting of collections of data and operations over these data.

Furthermore, programs are specified by means of logical axioms in a logical theory, typically containing equality. Algebraic specification languages typically follow the property-oriented style of specification, and they do not provide an explicit representation of concurrency. One of the most well-known languages that fall into this class is OBJ (Futatsugi et al., 1985; Goguen et al., 1988), a functional and object-oriented specification language. It is also an executable language, in that OBJ programs can be evaluated by interpreting the equations of the language as rewriting rules. Another example of an algebraic specification language is CASL (Common Algebraic Specification Language) (Mossakowski et al., 2003, 2008). It is based on first-order logic with induction. Several extensions of the base language have been designed, including a higher-order extension, an extension to model concurrency, and extensions based on temporal and modal logics.

5.6.2 Model-Based Specification Languages

Languages that fall into this class follow the model-based style of specification; that is, they provide an explicit mathematical model of the system state and the corresponding operations. As for the algebraic approach, typically no explicit model of concurrency is provided, although language extensions may be provided for this purpose.

The most well-known specification languages in this class are Z and VDM. VDM (Vienna Development Method) (Jones, 1990) specifications focus on operations. Each operation is specified by means of its state, a set of preconditions and a set of postconditions. A state consists of a set of external variables. Preconditions state the assumptions that must hold for the given operation to be executed, whereas postconditions state the logical conditions that must hold after the operation has been executed. Operations can be organized into hierarchical specifications. An extension of VDM also supports object-oriented and concurrent systems.

Z (Spivey, 1992) is a formal specification language based on typed set theory and first-order predicate logic. A specification in Z is given via a set of schemas, each schema providing a definition of a set of entities and the interrelationships between them. Each schema comprises a signature and a predicate section. The signature section declares the entities being modeled and their types, whereas the predicate section contains the theory that axiomatizes the relationships between the entities declared in the signature. As for VDM, a complete Z specification is typically based on a hierarchy of schemas.

Related to the Z notation are the B language and method (Abrial et al., 1991; Abrial, 1996). The B method is a comprehensive formal approach to specification and development. In B, a system is modeled as a set of interdependent abstract machines, a formalism in the same style as state-based specifications available in VDM and Z, and specifications are based on an object-oriented approach. Compared to Z, B is more focused on refinement rather than just system specification.

Other examples of model-based specification languages include Larch (Guttag and Horning, 1993) and Alloy (Jackson, 2006).

5.6.3 Process Algebras and Calculi

Process algebras and calculi (Baeten, 2005) include a set of logical formalisms used to describe concurrent and distributed systems, and explicitly model interactions (e.g., parallel and sequential composition of processes), inter-process communication, and synchronization. System behavior is described using an axiomatic approach, and formalized in terms of algebraic laws. The basic laws involve properties of the composition operators, for instance commutativity and associativity of parallel composition, and are common to all process algebras. Additional laws, such as those connecting parallel composition with the other operators, may differ depending on the specific process algebra. Languages in this class differ in the choice of semantic domain. Some languages focus on system states, some focus on events, and some deal with both. Central to all process calculi is a notion of *bisimulation equivalence*, which permits formal reasoning about equivalences between processes in terms of behaviors. The most famous examples of languages in this class are CSP (Hoare, 1985) (Communicating Sequential Processes), CCS (Calculus of Communicating Systems) (Milner, 1980), and ACP (Algebra of Communicating Processes) (Bergstra and Klop, 1984). More recent languages include π-*calculus* (Milner et al., 1992), with its several variants and extensions, and *ambient calculus* (Cardelli and Gordon, 2000), which focus on networks of processes with dynamic reconfiguration. The semantics of these languages are typically defined in terms of sequences, trees, or partial orders of events. For instance, in CCS, system behavior is interpreted over a set of trees of states and associated events.

Another language falling into this class is LOTOS (Brinksma, 1989), a specification language targeted to protocol specification and based on a language for specifying processes similar to CCS and CSP. This language provides support for reasoning about the combination of processes and data.

5.6.4 Logic-Based Languages

We group in this class different formalisms that use logic, in a broad sense, to describe systems and system properties. Specification languages based on propositional or first-order logic that are supported by theorem provers (compare Section 5.7.2) fall into this class, as do many other languages targeted at specifying timing behavior, such as various forms of temporal (Pnueli, 1981; Emerson, 1990) and interval logics (Moszkowski and Manna, 1983). *Duration calculus* (Chaochen et al., 1991) is a form of interval logic targeted at specifying real-time behavior. Temporal logic can be used in the property-oriented style to specify the behavior of concurrent and distributed systems. Variants of temporal logics provide different types of temporal operators, to reason about past, present, and future states and events. We present two variants of temporal logic, namely Linear Temporal Logic (LTL) and Computation Tree Logic (CTL), in Section 5.6.6.

Another class of logic-based languages includes various forms of state machines. State machines come in different flavors. They generally formalize system behavior in terms of sets of states, transitions between states, inputs, and outputs (actions). A broad classification distinguishes **Moore machines**, in which the outputs depend only on the state, from **Mealy machines**, in which the outputs depends on state and input. We discuss finite-state machines in Section 5.6.5. A further distinction is made between *deterministic* and *nondeterministic* machines; in the latter case, there may be more than one transition for each possible state, with a given input. Statecharts (Harel, 1987) are a variant of state machines that allows, among other things, explicit modeling of concurrent execution and communication between different machines, nested states, and hierarchical decomposition.

An alternative to describe system behavior, in particular program execution for software systems, uses **flowcharts**. A flowchart describes a program by explicitly modeling program statements and the control flow connecting the statements. Classical formalisms for reasoning about programs described as flowcharts include Hoare's approach (Hoare, 1969) and Dijkstra's weakest precondition (Dijkstra, 1976). In Hoare's approach, properties of programs are specified in terms of sentences of the form $P\{S\}Q$, where S is a program fragment, P is called *precondition*, and Q *postcondition*, with the intuitive meaning that executing S in a state satisfying the precondition P ends up, if the program terminates, in a state satisfying the postcondition Q. Reasoning about programs is achieved through a set of inference rules based on sentences.

Finally, Petri Nets (Petri, 1966; Peterson, 1981) are a well-known formalism for specifying concurrent systems in terms of the data flow through a network. In its basic form, Petri Nets consist of *places*, *transitions*, and (input and output) *arcs* connecting transitions with places. Places may contain *tokens*, which can be moved from one place to another as a result of transitions. Transitions are guarded by preconditions over the number of available tokens. The current state of the system being modeled is given by a *marking* that specifies the number of tokens in each place. Several variants of the basic formalism exist (e.g., timed and stochastic variants).

In the remainder of the chapter we focus on formal models given as finite-state machines, and represented as **Kripke structures** (also called **Kripke models**). Kripke structures are a traditional formalism to describe reactive systems, that is, nonterminating systems with infinite behaviors such as communication protocols, operating systems, or hardware circuits, and to formalize their dynamic evolution over time. Kripke structures form the basis for model checking, an automated formal verification technique that works by performing an exhaustive search over the state space of a given system. In model checking, the system is formalized as a Kripke structure, and system properties are typically expressed in a formalism based on temporal logic. Model checking is discussed in detail in Section 5.7.3.

5.6.5 State Transition Systems

A state transition system is defined as follows.

Definition 5.1 (State Transition System) Let $\mathcal{P}$ be a set of propositions. A state transition system is a tuple $\langle \mathcal{S}, \mathcal{I}, \mathcal{R}, \mathcal{L} \rangle$ where

- $\mathcal{S}$ is a finite set of states.
- $\mathcal{I} \subseteq \mathcal{S}$ is the set of initial states.
- $\mathcal{R} \subseteq \mathcal{S} \times \mathcal{S}$ is the transition relation.
- $\mathcal{L} : \mathcal{S} \longrightarrow 2^{\mathcal{P}}$ is the labeling function.

The transition relation specifies the possible transitions of the system. For technical reasons, we require it to be total; that is, for each state there exists at least one successor state. We notice that transition systems can model nondeterministic behavior by allowing states to have multiple successors. From the state transition system it is possible to extract the **state transition graph**, also known as a *Kripke structure*, that is, a graph describing the transitions from state to state.

Example 5.1 Let $\mathcal{M} = \langle \mathcal{S}, \mathcal{I}, \mathcal{R}, \mathcal{L} \rangle$ be a state transition system, where $\mathcal{S} = \{A, B, C, D\}$, $\mathcal{I} = \{A\}$, and $\mathcal{R} = \{(A, B), (B, C), (B, D), (C, D), (C, C), (D, D)\}$. The corresponding state transition graph is shown in Figure 5.1, where the initial state is in gray. ∎

A **trace** in a Kripke structure is obtained starting from an initial state in $\mathcal{I}$ and then repeatedly appending states that are reachable through $\mathcal{R}$. Formally, a trace can be defined as follows.

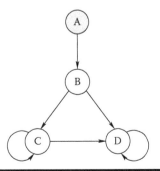

Figure 5.1   **An example Kripke structure.**

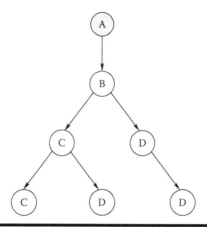

Figure 5.2   The unwinding of the Kripke structure in Figure 5.1.

Definition 5.2 (Trace)   Let $\mathcal{M} = \langle \mathcal{S}, \mathcal{I}, \mathcal{R}, \mathcal{L} \rangle$ be a Kripke structure. A trace for $\mathcal{M}$ is a sequence $s_0, s_1, \ldots, s_k$ such that $s_i \in \mathcal{S}$, $s_0 \in \mathcal{I}$ and $(s_{i-1}, s_i) \in \mathcal{R}$ for $i = 1 \ldots k$.

Given the totality of the transition relation, Kripke structures admit infinite traces, also called **paths** (formally, we may admit $k = \infty$ in the previous definition in order to accommodate infinite traces). A path in a Kripke model is therefore an infinite sequence of states $\sigma = s_0, s_1, s_2, \ldots$ satisfying the conditions of Definition 5.2. We say that a state s is reachable if there exists a path from an initial state to s. In general, because a state can have more than one successor, the Kripke structure can be thought of as unwinding into an infinite tree, representing all the possible paths of the system starting from the initial states.

Example 5.2   Figure 5.2 shows the unwinding of the Kripke structure in Figure 5.1 starting from the initial state A. ∎

The labeling function $\mathcal{L}$ maps each state s of the Kripke structure onto the set of propositions from $\mathcal{P}$ that hold in that state (the association is not shown in the previous example). We write $s \models p$ to indicate that a proposition p holds in a state s.

Kripke structures are usually presented using a structured programming language that provides constructs to represent their components. In this book we use the NuSMV language (McMillan, 1993; NuSMV, 2009; Cimatti et al., 2000, 2002) (see Appendix A for more information) as an exemplification. In NuSMV, each component can be specified by means of

■ State variables, which determine the state space $\mathcal{S}$ and the labeling $\mathcal{L}$
■ Initial values for the state variables, which determine the set of initial states $\mathcal{I}$
■ A set of instructions, which determine the transition relation $\mathcal{R}$

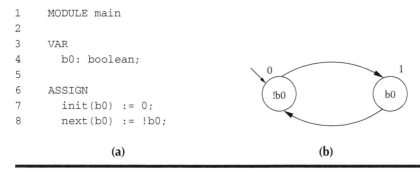

```
1    MODULE main
2
3    VAR
4       b0: boolean;
5
6    ASSIGN
7       init(b0) := 0;
8       next(b0) := !b0;
```

(a)   **(b)**

Figure 5.3 **(a) A simple NuSMV program with one variable; and (b) the state space associated with the NuSMV program in (a). (Opera by Fondazione Bruno Kessler released under a Creative Commons Attribution 3.0 United States License.)**

Example 5.3 A simple NuSMV program is shown in Figure 5.3. It consists of a variable declaration section (lines 3 and 4) and an assignment section (lines 6 to 8), whose purpose is to define the initial and next values of the variables. The program declares a Boolean variable b0 (line 4). Its initial value is set to 0 (line 7), whereas the transition relation is such that the next value of b0 is defined to be the negation of its current value (line 8).

The state space associated with the NuSMV program is given by the Kripke structure $\mathcal{M}$ shown in Figure 5.3b, where the initial state is pointed to by an arrow, and the two states (called 0 and 1) are labeled with the value of the atomic proposition b0. Symbolically, $\mathcal{P} = \{b0\}$, $\mathcal{M} = \langle \mathcal{S}, \mathcal{I}, \mathcal{R}, \mathcal{L} \rangle$, where $\mathcal{S} = \{0, 1\}$, $\mathcal{I} = \{0\}$ $\mathcal{R} = \{(0, 1), (1, 0)\}$, and $\mathcal{L} : \mathcal{S} \longrightarrow 2^{\mathcal{P}}$ is such that $\mathcal{L}(0) = \emptyset$ and $\mathcal{L}(1) = \{b0\}$. ■

In NuSMV, components can be combined either via **synchronous composition** or **asynchronous composition**. In synchronous composition (the default in NuSMV), components evolve in parallel, performing transitions simultaneously. Asynchronous composition is based on an interleaving model for the evolution of components (one component is selected to perform a transition at any given time). Synchronous and asynchronous composition can be used to model different types of systems. For instance, a sequential hardware circuit is a typical example of a synchronous system, whereas a communication protocol is naturally modeled using an asynchronous model of computation.

Example 5.4 We extend the NuSMV program of Example 5.3 by introducing an additional Boolean variable b1, which is left unconstrained, as in Figure 5.4. The corresponding state space is shown in Figure 5.4b. The new state space is obtained by computing the Cartesian product of the ranges of the two variables. It consists of the Kripke structure of Example 5.3, replicated twice (with solid arcs representing the corresponding transitions), and of additional transitions (shown with dashed arcs) that are induced by the new variable b1. ■

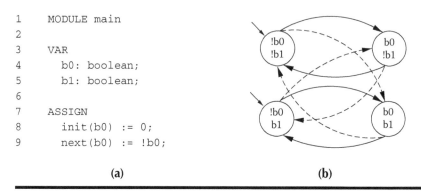

```
1     MODULE main
2
3     VAR
4       b0: boolean;
5       b1: boolean;
6
7     ASSIGN
8       init(b0) := 0;
9       next(b0) := !b0;
```

(a)   (b)

Figure 5.4 (a) A simple NuSMV program with two variables; and (b) the state space associated with the NuSMV program in (a). (Opera by Fondazione Bruno Kessler released under a Creative Commons Attribution 3.0 United States License.)

NuSMV provides some predefined data types, including Boolean, enumerations, and bounded integer variables. More details can be found in Appendix A.

5.6.6 Temporal Logic

Temporal logic (Pnueli, 1981; Emerson, 1990) can be used to express properties of reactive systems modeled as Kripke structures. System properties can be roughly classified into two different categories: **safety properties** and **liveness properties**. Safety properties express the fact that "nothing bad ever happens," whereas liveness properties state that "something desirable will eventually happen." For instance, the fact that it is never the case that two processes are running simultaneously inside their critical section is an example of a safety property for a communication protocol, whereas requiring that a process will sooner or later return control is an example of a liveness property. Safety properties can be refuted by showing a finite behavior, that is, a trace ending in a state satisfying p, where p models the *bad state* that should not be reached, as shown in Figure 5.5. On the other hand, liveness properties can be refuted by showing an infinite behavior, that is, a path along which p never holds,

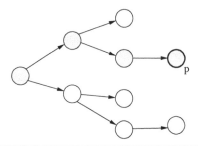

Figure 5.5 A finite trace refuting a safety property. (Opera by Fondazione Bruno Kessler released under a Creative Commons Attribution 3.0 United States License.)

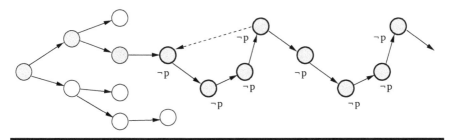

Figure 5.6 **An infinite trace refuting a liveness property. (Opera by Fondazione Bruno Kessler released under a Creative Commons Attribution 3.0 United States License.)**

where p models the desirable state that should be reached, as shown in Figure 5.6. In practice, in finite state systems an infinite path can be represented by a loop, as shown by the dashed arrow in Figure 5.6.

Traditionally, two temporal logics are most commonly used for model checking: Linear Temporal Logic (LTL) and Computation Tree Logic (CTL). A distinguishing feature of both logics is that they can be used to express properties over infinite behaviors, and that they do not make any explicit reference to time (time is modeled as an abstract transition step). LTL and CTL have incomparable expressive power, and they use a different model of time: LTL employs a linear model of time, whereas CTL is based on a branching model. A comparison between LTL and CTL, and a discussion about their relative merits, can be found in Vardi (2001).

Semantically, CTL is interpreted over the computation tree obtained as unwinding of the Kripke structure, as illustrated in Figure 5.2 (see Example 5.2). On the other hand, LTL is interpreted over the set of linear paths of the Kripke structure.

Example 5.5 The set of linear paths associated with the Kripke structure in Figure 5.1 is shown in Figure 5.7. ∎

We present below, in a mostly informal way, the syntax and semantics of propositional LTL and CTL. More details can be found in (Emerson, 1990; Clarke et al., 2000; Baier and Katoen, 2008). In the following we assume that a Kripke structure $\mathcal{M} = \langle \mathcal{S}, \mathcal{I}, \mathcal{R}, \mathcal{L} \rangle$ is given. We remind the reader that a proposition p is true in a state s_i if $p \in \mathcal{L}(s_i)$, where $\mathcal{L}$ is the labeling function.

Linear Temporal Logic. An LTL formula is evaluated over a linear path, that is, a sequence of states $s_0, s_1, s_2, \ldots, s_i, s_{i+1}, \ldots$. LTL provides the following temporal operators:

■ "Finally" (or "future"): Fp is true in a state s_i if and only if p is true in some state s_j, with $j \geq i$.
■ "Globally" (or "always"): Gp is true in a state s_i if and only if p is true in all states s_j, with $j \geq i$.

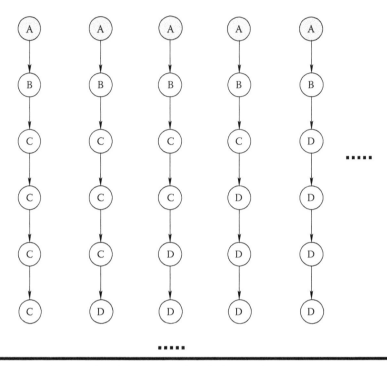

Figure 5.7   The set of paths associated with the Kripke structure in Figure 5.1.

- ◼ "Next": Xp is true in a state s_i if and only if p is true in the state s_{i+1}.
- ◼ "Until": pUq is true in a state s_i if and only if
    - – q is true in some state s_j with $j \geq i$.
    - – p is true in all states s_k such that $i \leq k < j$.

The semantics of linear temporal logic can be represented pictorially as in Figure 5.8.

Additional temporal operators can be defined on the basis of the operators described above. Examples of properties that can be expressed in LTL are as follows. An example of a safety property is

$$G(nr\_fail \leq 2) \rightarrow G(\neg bad) \tag{5.1}$$

which expresses the fact that, for every path, if the number of failures is less than or equal to 2, then it is never the case that the *bad* state is reached. An example of a liveness property is

$$G(input \rightarrow F\ output) \tag{5.2}$$

which states that on every path, an *input* will be eventually followed by one *output*.

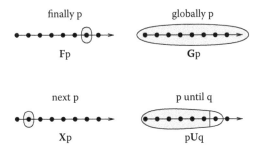

Computation Tree Logic. As noted, CTL is based on a branching model of time, and CTL formulae are interpreted over computation trees. CTL provides the same temporal operators as LTL, that is, F, G, X, U. In CTL, every temporal operator must be preceded by a *path quantifier*, a universal quantifier (A), or an existential quantifier (E). As a consequence, in CTL, quantifiers and temporal operators come in pairs. The universal quantifier originates the so-called *universal modalities* (AF, AG, AX, AU), whereas the existential quantifier originates the *existential modalities* (EF, EG, EX, EU). Universal modalities express the fact that a temporal formula must be true in *all* the paths starting from the current state, whereas existential modalities constrain a temporal formula to being true in *some* of the paths starting from the current state. The semantics of computation tree logic can be represented pictorially as in Figure 5.9.

Examples of properties that can be expressed in CTL are as follows. An example of a safety property is

$$AG\neg(Critical_1 \wedge Critical_2) \tag{5.3}$$

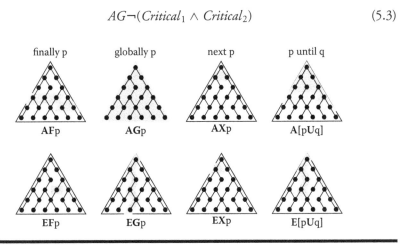

which expresses the problem of mutual exclusion for two processes executing their critical sections *Critical₁* and *Critical₂*. An example of a liveness property is

$$AG(Turn \rightarrow AF\ Critical) \tag{5.4}$$

stating that whenever a process gets the turn, it will be eventually allowed to execute its critical section.

In temporal model checking, **fairness conditions** (Clarke et al., 2000; Baier and Katoen, 2008) can be considered, in order to restrict the evaluation of temporal properties to the Kripke models satisfying some fairness constraint. Intuitively, the goal of fairness conditions is to rule out "unwanted" behaviors. For instance, if we want to prove a liveness property for a mutual exclusion protocol, stating that every process is eventually allowed to enter its critical section, then we can add a fairness constraint guaranteeing that every process will be scheduled an infinite number of times for execution. Formally, the set of fairness conditions is a set $F = \{f_1, \ldots, f_n\}$, where $f_i \subseteq S$ is a set of states. In the presence of fairness conditions, temporal formulae are evaluated only on **fair paths**, that is, paths such that at least one state for each f_i occurs in the path an infinite number of times.

PSL. The Property Specification Language (PSL) is a high-level, powerful language for expressing temporal requirements, developed by Accellera (Accellera, 2009) and standardized in IEEE 1850 (IEEE 1850, 2009). Its adoption is supported by the PSL/Sugar Consortium (PSL, 2009). The language originated in the hardware domain as a means to provide hardware designers with a common and interoperable language to formalize hardware specifications, and it comes in different flavors (it can be used in combination with hardware design languages such as VHDL or Verilog). PSL is based on IBM's Sugar language, which was originally conceived in order to provide syntactic sugar for CTL temporal logic. It was later extended to incorporate linear-time semantics and several additional features.

PSL is currently a powerful language, in that it is a superset of both CTL and LTL temporal logics, providing additional features such as the use of regular expressions and parameterized operators. The following example is the translation into PSL of Equation (5.4):

$$always(Turn \rightarrow eventually!\ Critical) \tag{5.5}$$

PSL admits powerful regular expressions such as, for example, $\{A;\ B[+];\ C[*2]\}$. Regular expressions are evaluated over finite paths. A path satisfying the previous expression is such that: A holds in the first time step; B holds for $n \geq 1$ time steps, starting with the second time step; finally, the remaining part of the path can be split into two parts, such that in both of them C holds.

We provide additional examples of PSL when discussing requirements validation in Section 5.7.4.

5.7 Formal Methods for Verification and Validation

In this section we discuss the application of formal methods for system verification and validation. Although not the main focus of this book, we first provide a short overview of traditional verification methodologies such as testing and simulation. Then we briefly discuss theorem proving techniques. Finally, we present in greater detail model checking, and we discuss the role of model checking in requirements validation and verification.

We now exemplify the role of formal methods techniques for verification and validation in the development process. The techniques that we present in this section are not specific to safety assessment. They can be used to prove different kinds of system properties, for instance functional correctness, reliability, and safety properties. The application of formal methods for the generation of artifacts that are specific to safety assessment is discussed in Section 5.8.

5.7.1 Testing and Simulation

Testing (Beizer, 1990; Beizer, 1995; Kaner et al., 1993; Myers, 2004) and simulation techniques are commonly used for both system verification and validation. Technically, they can be described as the process of evaluating the behavior of a given system (or a model of the system) by stimulating it with given inputs, and checking whether the produced outputs are conformant with the expected ones. This simple definition contains some implicit assumptions that need to hold in order for testing to be applicable. First, the system or model under test must be executable, in some form, either statically (e.g., by symbolic execution of a model of it) or dynamically (e.g,. by operation of the real system). Second, there must be a so-called **test oracle**, that is, a way to compare the outputs of testing with the expected outputs, in order to ensure that the system behaves as it is supposed to behave. A test oracle can be typically derived from the system requirements.

We use the term **simulation** to refer to the activity of testing involving a model of the system (rather than the system itself) or a model of the operating environment in which the system is supposed to work, or both. In the latter case, the term **environmental simulation** is often used. There are several reasons why simulation can be used instead of testing. System simulation can be carried out on a model of the system, when the final system is not available, for instance in the earlier stages of system design or development. Environmental simulation must be performed when testing a system in its intended operational environment is impossible or inconvenient (e.g., testing of space mission vehicles), or dangerous for safety reasons (e.g., testing of a shutdown system for a nuclear power plant). In other cases, an environmental simulator can be used to simulate a broader range of environmental situations than those typically reproducible in the real environment. There are a number of issues related to system or environmental simulation. In general, care must be taken to ensure that an environmental simulator is a good reproduction of the real operating environment. Moreover, for system simulation it may be problematic or impossible to test some properties (e.g., real-time properties) on a model of the system. In the remainder of this section, we disregard the issues related to simulation and focus our attention on testing. Most of the issues we discuss, however, are also applicable to simulation.

Forms of Testing. Depending on the way the tests are executed, traditionally two different forms of testing are distinguished, namely **static testing** and **dynamic testing**.

Static testing consists of investigating the system properties without actually operating the system. Static testing may assume different forms, for instance, static analysis of the code or code inspection in the case of software products. Dynamic testing involves the actual execution of the system, for instance, code execution in the case of software. Depending on the opacity of the system being tested, we further distinguish **black-box testing** from **white-box testing**. In black-box testing, the tester has no knowledge about the internal implementation of the system being tested. Typically, black-box testing is used to evaluate the functional correctness of a given system, independently of its internal implementation. It is often applied to evaluate the overall system, although individual components can also be evaluated in this way. Black-box testing facilitates independence of the tester from the evaluator, as the tester need not know the internals of the system under test. Black-box testing is sometimes called **requirements-based testing** because it relies on the system specification. White-box testing, on the other hand, is typically performed by test engineers who have complete knowledge of the internal implementation of a given system and can use this knowledge to guide the testing activity. Typically, static testing requires a white-box approach, whereas dynamic testing may be performed using both white-box or black-box techniques. With respect to the different forms of testing performed in the development life cycle (compare Chapter 4), the white-box or black-box approach may or may not be feasible. For example, both white-box and black-box approaches can be done for integration testing, whereas acceptance testing is typically performed using a black-box methodology. In the remainder of this section, we discuss static testing and dynamic testing in more detail.

Static Testing. In this book we discuss static testing in a very broad sense. We group under this name a plethora of techniques, based on formal verification, that are used to evaluate the behavior and characteristics of a given system without actually operating it. Some of these techniques, such as those involving code inspection, can be performed manually, whereas other techniques, such as those analyzing the control or data flow of programs, can be tool supported. Below we list some of the most popular static testing techniques.

■ **Walk-throughs or design reviews** are a form of code inspection, typically performed using peer review by a set of testing engineers.

- **Fagan inspections** refer to a systematic methodology to find defects and omissions in the development process, which is performed in different stages, namely planning, overview, preparation, inspection, rework, and follow-up. For each activity, entry and exit criteria are defined, and Fagan inspections are used to ensure that the development process complies with the exit criteria for each activity.
- **Formal proofs** include, in a very broad sense, most techniques, based on formal methods, that are used to prove properties of parts of the design or the implementation of a given system. Theorem proving techniques (discussed in Section 5.7.2) and model checking techniques (discussed in Section 5.7.3) can be considered instances of this class.
- **Type analysis** statically analyzes the type information of a software program, for instance, in order to ensure that type errors may not arise at run-time.
- **Control flow analysis** analyzes the control structure of a software program in order to detect problems in the program structure, such as code that is never executed or infinite loops. The control flow is formalized as a graph in which the nodes correspond to specific points in the code and the edges represent different flows.
- **Data flow analysis** analyzes the flow of data within a software program. Each operation on the data is analyzed, and the data flow is compared with the expected one. Data flow analysis can exploit different forms of static analysis techniques; for instance, abstract interpretation can be used to abstract the data values.
- **Symbolic execution** consists of running a given program using symbolic inputs in place of actual values. The results of operations on data are algebraic expressions, instead of values. The objective of symbolic execution is to compare the symbolic result of execution with the expected outputs of the program. Symbolic execution can exploit different forms of static analysis techniques (e.g., partial evaluation).
- **Pointer analysis** is a static technique that analyzes how a given program accesses to pointers or heap references, to determine issues such as memory leaks or wrong dereferences.
- **Shape analysis** is a generalization of pointer analysis. It statically analyzes a program to determine information about the heap-allocated data structures that the program manipulates and their shape (e.g., tree-like, or arbitrary graph). The results can be used to understand or verify programs, for instance, to verify that a program does not reference null or dangling pointers, and to prevent memory leaks. Shape analysis also produces valuable information for debugging, compile-time garbage collection, instruction scheduling, and parallelization.

Dynamic Testing. Dynamic testing consists of analyzing the response of a given system to a set of predetermined inputs, by operating the system in its operational environment (or a simulation of it). For software components, it consists of executing the system code.

Depending on the testing techniques, that is, the criteria used to devise the set of tests to be carried out, we categorize dynamic testing as follows:

- **Behavioral testing** is intended to evaluate the behavior of the system under test from the perspective of an external user. It is an example of black-box testing. It can be further distinguished into the following categories:
  - **Functional testing** evaluates the compliance with the functional requirements.
  - **Nonfunctional testing** evaluates nonfunctional characteristics of the system such as its performance, reliability, or safety.
- **Structural testing** is intended to test the various routines and different execution paths within a system. It requires knowledge of the internal implementation of the system, and therefore it is an example of white-box testing.
- **Random testing** involves a random choice of the test cases to be tried on the system under test. The purpose of random testing is to uncover faults that have not been uncovered by the previous techniques. Test cases can also be chosen according to a given probability distribution.

Test Coverage. Effective and high-quality testing requires a good choice of the test cases to be run. Each test case consists of the set of inputs to be applied to the system, called the **test vector**; possibly a set of preconditions that must be satisfied to execute the test; and the set of expected outputs. For all but the simplest systems, exhaustive testing is practically impossible, given the huge amount of applicable test vectors. For this reason, the set of test cases to be run must be carefully chosen in a way so as to maximize the probability of uncovering faults. The quality of the test cases is typically evaluated in terms of the notion of **test coverage** (the testing activity that is planned according to this notion of coverage is called **coverage-based testing**). A naive way to measure coverage would be to measure the percentage of test cases that are executed with respect to the number of possible test vectors. However, for most systems, the number of tests that can be run in practice would be a very small fraction of the total number of test vectors. Moreover, this notion of coverage does not consider the significance of test cases. For these reasons, different notions of coverage are typically used. For instance, test cases may be chosen depending on a particular fault model; that is, test cases are chosen in order to stimulate the system under test under the hypothesis that it may fail in some predefined ways. In this view, it is important to have good fault modes, in order not to miss important faults. An alternative notion of coverage that is applicable to black-box testing is called **requirements-based coverage**, and it consists of evaluating the percentage of functions within the requirements specification that are analyzed.

In the remainder of this section we discuss in more detail some of the most well-known testing techniques and their respective coverage criteria. In particular, some traditional testing strategies for functional analysis are the following ones:

- **Equivalence partitioning** requires the set of test vectors to be partitioned into equivalence classes that are expected to have similar (qualitative) behaviors by looking at the system from the outside (black-box view). One test case is then chosen and executed as a representative of each class. The assumption underlying this technique is that if a test passes for one specific value in a class, all the tests for the remaining values would be successful.
- **Boundary analysis** forces the set of test vectors to be chosen at the extremes (boundary) of each equivalence class. Tests are performed for values at either side of the boundary. The idea underlying this technique is to uncover faults that are due to "corner cases." A straightforward notion of coverage for both equivalence and boundary analysis measures the number of classes that have been tested.
- **State-transition testing** consists of stimulating each transition in the system, which can be seen as a finite-state machine. Full coverage would correspond to stimulating all transitions.

Structural testing includes several techniques, typically based on the data flow or on the control flow of the system under test. The idea of structural testing is to exercise all the possible execution paths within the system. The corresponding notion of coverage goes under the name of **structure-based coverage**. For instance, by considering the control-flow, the most well-known coverage measures are the following:

- **Statement coverage** measures the portion of statements that is actually executed during a test campaign. When using this technique, full coverage ensures that all instructions in the software have been executed at least once during a test. In practice, statement coverage is often considered a minimal requirement for coverage in software testing.
- **Branch coverage** measures, instead, coverage of execution paths. The goal here is to cover all possible combinations of conditional statements so that the test verifies the behavior of concatenation of interesting events. Thus, when using branch coverage, all possible combinations of branches (typically associated to Boolean conditions—e.g., the condition in an "if" statement) are computed, and full coverage is achieved when test cases verify all such combinations. This is more complex than statement coverage; for instance, in the piece of code shown in Figure 5.10, full statement coverage can be achieved with two test cases, while full branch coverage needs four test cases. In general, full branch coverage requires a number of test cases that are exponential in the number of branches, and hence often unrealistic in practice.
- **Call-graph coverage** forces each possible invocation tree of the subroutines of a given program to be exercised.

In some case, it may be impossible or inconvenient to achieve a full coverage. Nevertheless, standards may recommend or require specific **test adequacy** criteria

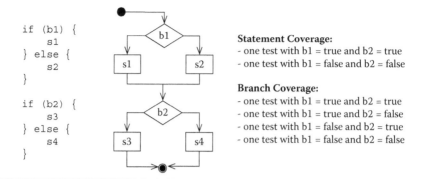

Figure 5.10 **Coverage example.**

in order to evaluate the effectiveness and quality of the testing activity. Such criteria may prescribe the use of specific testing techniques for particular testing phases and the level of coverage that should be targeted. Adequacy criteria typically depend on the integrity level of the particular application at hand. Sometimes, it is useful to plan the testing activity in a way that prioritizes the most important tests when the time that can be devoted to testing is limited.

5.7.2 Theorem Proving

Theorem proving is a verification technique that assumes that a system and the properties to be verified are expressed in a common language, based on mathematical logic. A theorem prover is based on a formal system, which consists of a set of axioms and a set of inference rules, that can be used to derive new theorems. Verifying a property amounts to generating a proof, where each deduction step in the proof is an instance of one of the inference rules, applied to axioms or previously proved theorems. Generation of lemmas, that is, intermediate proofs, can be used to aid the construction of the proof.

Theorem provers can be classified depending on the language they support. For instance, many theorem provers are based on *first-order logic*, or proper subsets of it. Some others support more complex theories such as different subsets of higher-order or modal logics. For all but the simplest logical systems, the problem of proving that a logical formula is a theorem is undecidable in general. Under suitable hypotheses (no proper axioms), by Gödel's completeness theorem, first-order predicate calculus is recursively enumerable, which means that any theorem, with unbounded resources, can be eventually proved valid, whereas invalid formulae cannot always be recognized. First-order predicate calculi with proper axioms may admit valid statements that are not provable in the theory, by Gödel's incompleteness theorems. Despite these theoretical limitations, theorem provers can still be used in practice to solve interesting problems, even in logical fragments that are undecidable.

Different theorem provers are based on different forms of inference rules (Gallier, 1986). Moreover, theorem provers typically use some form of mathematical or structural induction to reason over infinite domains. Some theorem provers use hybrid techniques; for instance, some theorem provers combine theorem proving capabilities with model checking techniques (compare Section 5.7.3). Some theorem provers are general-purpose, whereas others are targeted at the verification of specific problems.

Theorem provers are classified by the amount of automation in constructing proofs they provide (in the following we focus on machine-assisted theorem proving, although proofs can also be constructed by hand). They range from **automated theorem provers**, which run fully automatically, to **interactive theorem provers**, which require human guidance in providing hints to produce the proof. At the other extreme, **proof checkers** do not provide any proof construction capabilities; they can be used to verify that an existing proof for a theorem is a valid proof. Although proof checkers are less powerful than theorem provers, it may be easier and more effective (e.g., for certification purposes) to use a proof checker as an independent tool for validating a proof found by a theorem prover.

Examples of automated theorem provers are E (E, 2009); Otter (Otter, 2009) and its successor Prover9 (Prover9, 2009); Setheo (Setheo, 2009); SPASS (SPASS, 2009); and Vampire (Vampire, 2009). Among interactive theorem provers, we mention ACL2 (ACL2, 2009); HOL (HOL, 2009); PVS (PVS, 2009); and STeP (STeP, 2009). PVS and STeP also provide some model checking capabilities. Finally, proof checkers include Coq (Coq, 2009) and NuPRL (NuPRL, 2009).

5.7.3 Model Checking

Model checking (Clarke et al., 2000; Baier and Katoen, 2008; Grumberg and Veith, 2008) is a formal verification technique widely used to complement classical techniques such as simulation and testing. In particular, while testing and simulation may only verify a limited portion of the possible behaviors of complex systems, model checking provides a formal guarantee that some given specification is obeyed.

In model checking, the system under verification is modeled as a state transition system, and the specifications are expressed as temporal logic formulae that express constraints over the system dynamics. Model checking then consists of exhaustively exploring every possible system behavior to check automatically that the specifications are satisfied. In the case of finite models, termination is guaranteed. Very relevant for debugging purposes, when a specification is not satisfied, a counterexample is produced, witnessing the offending behavior of the system. Pictorially, a model checker can be represented as in Figure 5.11.

Formally, given a state transition system $\mathcal{M}$ and a temporal formula ϕ, model checking is the problem of deciding whether ϕ holds in $\mathcal{M}$, in symbols $\mathcal{M} \models \phi$.

The approach to model checking described in this section is also called **temporal model checking**, as it is based on expressing properties in temporal logic.

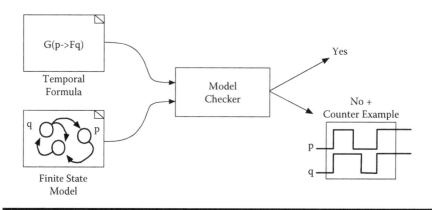

Figure 5.11 **A pictorial view of a model checker. (Opera by Fondazione Bruno Kessler released under a Creative Commons Attribution 3.0 United States License.)**

An alternative approach to model checking is based on using automata to specify both the system at hand and the property to be verified. In this approach, called **automata-based model checking**, verifying whether the system satisfies the property amounts to checking suitable properties of the corresponding automata (e.g., checking language inclusion). Vardi and Wolper (1986) have shown how the problem of temporal model checking can be reformulated in terms of automata, thus reconciling the two approaches. In the remainder of this chapter we focus on temporal model checking.

From Explicit-State to Symbolic Model Checking. In its simpler form, referred to as **explicit state**, model checking is based on the expansion and storage of individual states. The first model checking algorithms used an explicit representation of the Kripke structure as a labeled, directed graph. These techniques suffer from the so-called **state explosion problem**; that is, they need to explore and store the states of the state transition graph. In general, the Kripke structure may result as the combination of a number of components (e.g., the communicating processes in a protocol); hence its size may be exponential in the number of components.

A major breakthrough was enabled by the introduction of **symbolic model checking** (McMillan, 1993). The idea is to manipulate *sets of* states and transitions, using a logical formalism to represent the characteristic functions of such sets. Because a small logical formula can admit a large number of models, this results in many practical cases in a very compact representation that can be effectively manipulated.

In this view, each state is represented by an assignment to the propositions (variables) in $\mathcal{P}$, and a Kripke structure $\mathcal{M} = \langle \mathcal{S}, \mathcal{I}, \mathcal{R}, \mathcal{L} \rangle$ can be represented as follows. Without loss of generality, we assume that there exists a bijection between $\mathcal{S}$ and $2^{\mathcal{P}}$ (if the cardinality of $\mathcal{S}$ is not a power of 2, standard constructions can be applied to extend the Kripke structure). We represent the set of states $\mathcal{S}$

with a vector of Boolean variables (for non-Boolean variables, a Boolean encoding can be performed). We use $\underline{x}$ to denote the vector of such variables that we call *state variables*, and we assume a fixed correspondence between the propositions in $\mathcal{P}$ and the variables in $\underline{x}$. We write $\mathcal{I}(\underline{x})$ for the formula representing the initial states. To represent the transitions, we introduce a set of "next" variables $\underline{x}'$, used for the state resulting after the transition. A transition from s to s' is then represented as a truth assignment to the current and next variables. We use $\mathcal{R}(\underline{x}, \underline{x}')$ for the formula representing the transition relation expressed in terms of those variables.

Note that operations over sets of states can be represented by means of Boolean operators. For instance, intersection amounts to conjunction between the formulae representing the sets, union is represented by disjunction, complement is represented by negation, and projection is realized with quantification.

As an example, we show how the *image* and *preimage* operators can be encoded symbolically. The image operator, denoted *fwd_img*, computes the forward image of a set of states with respect to the transition relation of a Kripke structure $\mathcal{M}$; that is:

$$ fwd\_img(\mathcal{M}, Q) = \{s' \mid \exists s \in Q. \; \mathcal{R}(s, s')\} \tag{5.6} $$

where Q is a set of states. Symbolically, it can be encoded, using conjunction and existential quantification, as follows:

$$ fwd\_img(\mathcal{M}, Q(\underline{x})) = \exists \underline{x}. \; (Q(\underline{x}) \wedge \mathcal{R}(\underline{x}, \underline{x}')) \tag{5.7} $$

Similarly, the preimage of Q, computing the backward image of a set of states, is defined as

$$ bwd\_img(\mathcal{M}, Q) = \{s \mid \exists s' \in Q. \; \mathcal{R}(s, s')\} \tag{5.8} $$

Symbolically, it can be encoded as follows:

$$ bwd\_img(\mathcal{M}, Q(\underline{x})) = \exists \underline{x}'. \; (Q(\underline{x}') \wedge \mathcal{R}(\underline{x}, \underline{x}')) \tag{5.9} $$

Binary Decision Diagrams. Another key issue is the use of an efficient machinery to carry out the manipulation. Traditionally, Ordered Binary Decision Diagrams (OBDDs, or BDDs for short) (Bryant, 1992) have been extensively used for this purpose. BDDs are a representation for Boolean formulae, which is canonical once an order on the variables has been established. This allows equivalence checking in constant time. The basic set of theoretic operations on sets of states (intersection, union, projection) are given by logical operations on BDDs (such as conjunction, disjunction, and quantification). BDDs provide primitives to efficiently compute all these operations.

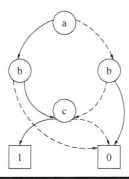

Figure 5.12   A BDD for the formula $(a \leftrightarrow b) \wedge c$.

Example 5.6   Figure 5.12 depicts the BDD for the Boolean formula $(a \leftrightarrow b) \wedge c$, using the variable ordering a, b, c. Solid lines represent "then" arcs (the corresponding variable must be considered positive), whereas dashed lines represent "else" arcs (the corresponding variable must be considered negative). Paths from the root to the node labeled "1" represent the satisfying assignments of the represented Boolean formula (e.g., $a \leftarrow 0$, $b \leftarrow 0$, $c \leftarrow 1$). ■

Efficient packages are available for BDD manipulation. Despite the worst-case complexity (e.g., certain classes of Boolean functions are proved not to have a polynomial-size BDD representation for any variable order), in practice it is possible to represent Kripke structures effectively using BDDs.

The use of BDDs makes it possible to verify very large systems (larger than 10^{20} states [Burch et al., 1992; McMillan, 1993; Burch et al., 1994]). Symbolic model checking has been successful in various fields, allowing the discovery of design bugs that were very difficult to highlight with traditional techniques. For instance, Clarke et al. (1993) discovered previously undetected and potential errors in the design of the cache coherence protocol described in the IEEE Futurebus+ Standard 896.1.1991, and the cache coherence protocol of the Scalable Coherent Interface, IEEE Standard 1596–1992, was verified (Dill et al., 1992), finding several errors.

Bounded Model Checking.   A more recent advance in the field of symbolic model checking goes under the name of **bounded model checking** (BMC) (Biere et al., 1999). In bounded model checking, the search for a trace that witnesses a property violation is limited to a certain bound, say k steps. Moreover, the problem is encoded into a propositional formula that is satisfiable if and only if a witness to the property violation exists. The formula is obtained by unwinding the symbolic description of the transition relation over time, as follows:

$$\mathcal{I}(\underline{x_0}) \wedge \mathcal{R}(\underline{x_0}, \underline{x_1}) \wedge \ldots \wedge \mathcal{R}(\underline{x_{k-1}}, \underline{x_k}) \tag{5.10}$$

where $\underline{x}_0, \ldots, \underline{x}_k$ are vectors of state variables whose assignments represent the states at the different steps. Additional constraints are used to limit such assignments to those that witness the violation of the property, and to impose a cyclic behavior when required.

Technically, the solution of a bounded model checking problem leverages the power of modern SAT solvers (Eèn and Sörensson, 2003), which in many practical cases are able to check the satisfiability of formulae with hundreds of thousands of variables and millions of clauses. In comparison to BDD-based algorithms, the advantages of SAT-based techniques are twofold. First, SAT-based algorithms have higher capacity; that is, they can deal with a larger number of variables. Second, SAT solvers have a high degree of automation and are less sensitive than BDDs to the specific parameters (e.g., variable ordering). On the other hand, SAT-based algorithms may be less effective than BDD-based ones in proving the correctness of specifications (as opposed to finding bugs) and may have problems when long counterexample traces are involved.

SAT-based technologies have been introduced in industrial settings to complement and sometimes replace BDD-based techniques. In addition, SAT has become the core of many other algorithms and approaches, such as inductive reasoning (e.g., Sheeran et al. (2000)), incremental bounded model checking (e.g., Heljanko et al. (2005)), and abstraction (e.g., Gupta et al. (2004)). Bounded model checking has also been incorporated in several model checkers. Verification techniques similar to bounded model checking have been adopted since the early 1990s by Gunnar Stålmarck and the associated company Prover Technology, based on a patented SAT solver. A survey of the recent developments can be found in Prasad et al. (2005).

Despite its recent introduction, bounded model checking is now widely accepted as an effective technique that complements BDD-based model checking. A typical methodology used in the industry today is to use both BMC and BDD-based model checkers as complementary methods. In some cases, both tools are run in parallel, and the first tool that terminates kills the other process. In other cases, BMC is used first to quickly find the more shallow bugs, and when this becomes too hard, an attempt to prove that the property is correct is being made with a BDD-based tool. In any case, it is clear that, together with the advancements in the more traditional BDD-based symbolic model checkers, formal verification of finite models has made a big step forward in the past few years.

A Short History. As mentioned in Section 5.7.3, the first model checkers used explicit state model checking techniques; that is, they represented the state space as a graph. Among them, we mention EMC (Clarke and Emerson, 1981; Clarke et al., 1986) and CÆSAR (Queille and Sifakis, 1982). Another example of explicit state model checker is SPIN (SPIN, 2009). The first symbolic model checker, called SMV, is due to McMillan (1993); it was the first model checker to use Binary Decision Diagrams. Since then, several symbolic model checkers have been designed and developed. Among them, we mention NuSMV (NuSMV, 2009) and VIS (VIS, 2009).

NuSMV (Cimatti et al., 2002) originates from a reengineering, reimplementation, and extension of SMV. Finally, STeP (STeP, 2009), the Stanford theorem prover, combines deductive capabilities with model checking.

Some model checkers can deal with more complex theories than finite-state machines. For example, KRONOS (Daws and Yovine, 1995) and UPPAAL (UPPAAL, 2009) can deal with timed systems, whereas HyTech (Alur et al., 1996) is a model checker for hybrid systems. Some model checkers have been extended to deal with probabilistic systems, among them PRISM (2009) and MRMC (2009).

Modern symbolic model checkers exploit advanced techniques that can significantly improve their performance and extend the applicability of model checking. Among these techniques, we mention the use of abstraction techniques, and the use of *divide and conquer* verification techniques such as compositional and *assume-guarantee* reasoning. Moreover, we mention **partial order reduction**, a technique to reduce the state space by exploiting the commutativity of concurrent executions in asynchronous systems; **symmetry reduction**, a technique that exploits symmetries in a given model to scale it down to an equivalent one; and model transformation techniques, such as minimization techniques based on bisimulation. For a thorough discussion of these and other techniques, and for a more extensive treatment of the theoretical background underlying model checking, we refer the interested reader to Clarke et al. (2000) and Baier and Katoen (2008).

In the remainder of this chapter we present the application of symbolic model checking from a practical perspective. In particular, we first exemplify the role of model checking for requirements validation in Section 5.7.4 and then discuss the use of model checking in formal verification in Section 5.7.5.

5.7.4 Using Model Checking for Requirements Validation

We start our discussion of the role of model checking in formal verification from the problem of validating a set of requirements. As mentioned in Section 5.2, this problem consists of ensuring that system requirements match end-user expectations. This is clearly a critical issue in system development. It is meaningless to formally prove that a product being developed conforms to a set of properties if these properties do not capture the design intent and do not correspond to what the customer had in mind. Furthermore, errors in the requirements manifest themselves late in the development process, and are likely to be very expensive to correct. Interestingly enough, practical experience in industry shows that flaws in the requirements engineering phase are responsible for about 50% of product defects and about 80% of the re-engineering effort once the defects have been revealed (Wiegers, 2001). For these reasons, it is clear that requirements quality in system development is of utmost importance. Moreover, formal methods have the potential to deal with this issue, in that they rely on formal specifications that eliminate ambiguities in the requirements and make them amenable to automatic verification. In this section we discuss the role of formal techniques, based on model checking, for improving requirements quality. The techniques we describe are implemented in the RAT tool (Pill et al., 2006; RAT, 2009). We use RAT to exemplify a typical process for quality control of a set of requirements.

The RAT tool provides complementary ways to assess the quality of requirements, helping designers in debugging and correcting specification errors. The first functionality is called **property simulation**. Simulating a set of properties allows the designer to explore the behaviors associated with each of the requirements. The RAT tool generates a set of traces that are representative of the requirements, and then allows the designer to generate alternative traces by adding constraints to the simulation (e.g., by setting one variable to assume a particular value at a given time instant—alternatively, at all time instants). The RAT tool can check if the newly produced traces conform to the requirements, and, if not, produce an alternative trace that does, while still satisfying the user constraints.

The second set of functionalities provided by the RAT tool is referred to as **property assurance**. It includes a few different functionalities. First, it allows us to check for *logical consistency*. Logical consistency can be intuitively described as "freedom from contradictions." In other words, a check is done in order to ensure that there are no properties mandating mutually incompatible behaviors. If the set of requirements is logically inconsistent, the tool is able to single out a subset of the requirements that is an explanation for the inconsistency. This allows the designer to pinpoint the problem in the specification. Second, it is possible to check whether the set of properties is *strict enough* to rule out unwanted behavior and *not too strict* to disallow for certain desirable behavior. These checks are performed by defining a set of *assertions* and a set of *possibilities*. Assertions are user-defined properties that are supposed to be logical consequences of the original sets of requirements, whereas possibilities are user-defined properties that are supposed to be consistent with (i.e., they are allowed by) the original set. Checking whether the set of requirements is strict enough amounts to verifying whether an assertion (describing the desired behaviors) is implied by the original set of requirements. If the property is not implied by the requirements, the tool can show a counterexample witnessing the violation of the property. Checking that the set of requirements is not too strict amounts to checking whether a possibility is compatible with the requirements. If it is compatible, the tool can generate a witness trace satisfying both the requirements and the possibility, whereas if it not compatible, it is possible to generate a subset of the requirements that prevent the possibility. The behaviors used for checking compatibility can be partial, in order to describe a wide class of compatible behaviors.

Whenever an anomaly, such as an inconsistency or an undesired behavior, is uncovered, the designer is given useful feedback information to fix the specification (e.g., an explanation for the inconsistency or a counterexample trace). The designer has then the possibility to correct the specification and repeat the analyses that have been already performed. A possible process to support the design engineering phase is described in Pill et al. (2006).

All the RAT functionalities described in this section rely on model checking techniques and can be carried out by dedicated formal verification algorithms (Pill et al., 2006). In particular, the RAT verification engine is based on automata-based model checking and bounded model checking (compare Section 5.7.3). For instance, consistency checking amounts to checking the emptiness of the language of a suitable automaton, whereas verifying that an assertion is a logical consequence of a set of properties is similar in spirit to traditional model checking, where a property is checked against a model. Here the considered set of properties plays the role of the model against which the property must be verified.

We now exemplify the requirements engineering phase with a practical example. The requirements are written in PSL (see Section 5.6.6), using an LTL-like syntax. Notice that in this syntax, the temporal operators G and F correspond to the operators *always* and *eventually!* of Section 5.6.6, respectively.

Example 5.7 We formalize the requirements for operation of an elevator. The elevator must service requests made by users at three different floors, numbered from one to three. Each floor is equipped with a button that can be pressed to request service.

We first define the following signals:

```
button: array 1..3 of boolean
floor: 1..3
door: {open,closed}
direction: {standing,moving}
service: array 1..3 of boolean
```

The variable `button[i]` is true whenever there is a request at floor i, `floor` indicates the floor at which the cabin is situated, `door` indicates whether the door cabin is open or closed, and `direction` indicates whether the cabin is standing or moving. Finally, `service[i]` is true if the elevator is granting service at floor i.

Based on these signals, we define the initial set of requirements as:

```
(R1) forall i in {1:3} :
     G (service[i] <-> (floor=i && door=open))
(R2) forall i in {1:3} : G(button[i] -> F(service[i]))
(R3) G (door=open -> direction=standing)
(R4) forall i in {1:3} :
     G(floor=i -> (next(floor) <= i + 1) && next(floor) >= i - 1)
(R5) forall i in {1:3} :
     G((direction=standing && floor = i) -> (next(floor) = i))
(R6) forall i in {1:3} :
     G((direction=moving && floor = i) -> (next(floor) != i))
(R7) forall i in {1:3} : G(button[i] -> [button[i] U service[i]])
(R8) forall i in {1:3} : G(service[i] -> next(!button[i]))
```

The requirements have the following meaning. Requirement (R1) formalizes the definition of service; it states that the elevator is servicing at floor i if the cabin is at floor *i* with the door open. Requirement (R2) states that any request is taken care of eventually. Requirement (R3) is a safety requirement; it states that it is always the case that if the door is open, then the cabin is standing (i.e., not moving). Requirements (R4) through (R6) constrain the way the floor can change: (R4) states that the elevator can move at most one floor at a time; (R5) states that if the cabin is standing, then the floor does not change; finally, (R6) states that if the elevator is moving, then the floor changes. Requirement (R7) states that once a request has been issued, it cannot be reset until the request has been serviced. Finally, requirement (R8) states that, once a request has been serviced, it is reset at the next step. Notice that we are using the universal quantifier *forall* to formalize requirements that hold for each floor *i* ranging in {1, 2, 3}.

Using the RAT tool, we first check whether the set of requirements is consistent, and we get a positive answer. Then, we add the following assertion:

```
(A1) forall i in {1:3} : G(service[i] -> direction=standing)
```

It is a safety requirement stating that whenever any request is being serviced, the cabin must be standing. Again, the RAT tool returns with a positive answer; that is, the assertion is a logical consequence of the set of requirements.

We then check the following assertion:

```
(A2) forall i in {1:3} : G(service[i] -> next(!service[i]))
```

In other words, we are stating that requests at the same floor do not come too close together; in particular, we cannot have two requests at floor *i* being serviced in consecutive time steps. By running RAT, we discover that this assertion is violated. RAT returns with the counterexample trace shown in Figure 5.13.

In fact, the counterexample shows us a more general reason for the violation. Specifically, our requirements allow the elevator to grant a service at floor *i* even though the corresponding button has not been pressed. As a consequence, it may also be the case that two consecutive services at the same floor are granted. To rule out this unwanted behavior, we add the following requirements:

```
(R9)  forall i in {1:3} : [!service[i] U button[i]]
(R10) forall i in {1:3} :
      G(service[i] -> next([!service[i] U button[i]]))
```

Requirement (R9) states that there can be no service at floor *i* before the corresponding button is pressed; this requirement constrains only the first request at floor *i*. Requirement (R10) states a similar constraint on further requests arriving after the first one; it states that whenever service is granted at floor *i*, there must be no further service at the same floor before the button is pressed again. By running the RAT tool, we can verify that assertion (A2) is now verified; that is, it is a logical consequence of requirements (R1) through (R10). We can also check that the resulting set of requirements is still consistent, and assertion (A1) is still valid.

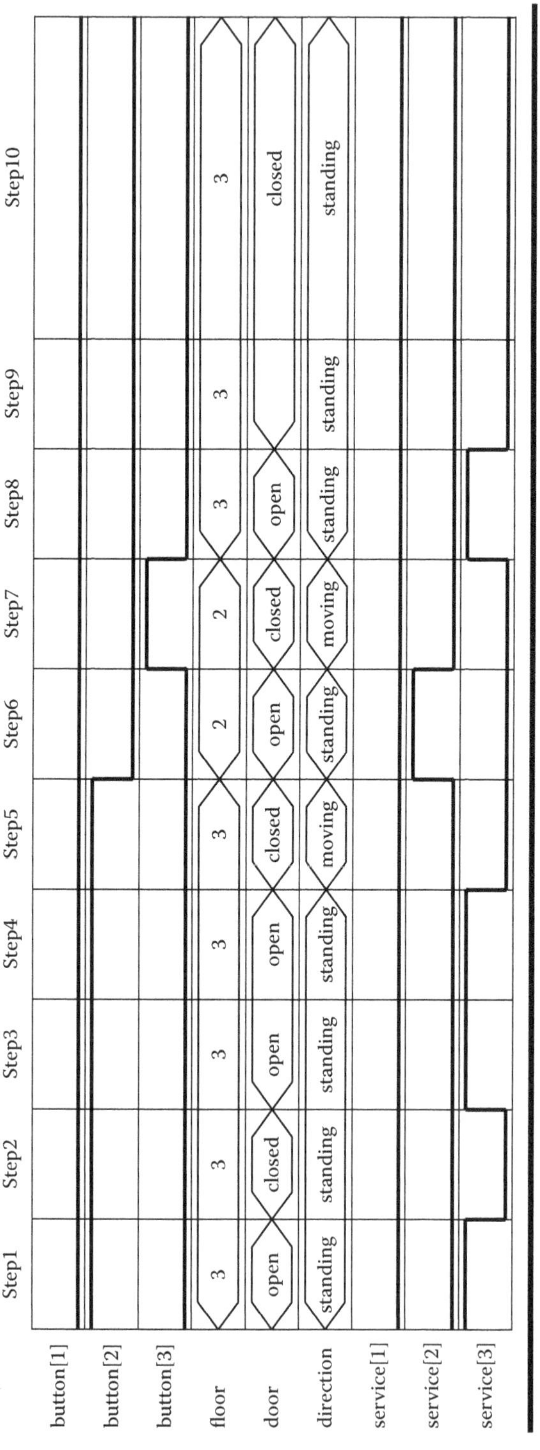

Figure 5.13 A counterexample trace for assertion (A2).

By running a few simulations of the requirements, we generate some traces that feature initial states when one or more buttons are pressed. We then want to investigate whether our requirements are compatible with a scenario in which initially no buttons are pressed. To this goal, we check the following possibility:

```
(P1) forall i in {1:3} : (!button[i])
```

The RAT tool verifies that this behavior is allowed, as desired. We generalize possibility (P1) as follows:

```
(P2) forall i in {1:3} : G(!button[i])
```

In other words, we look for a scenario in which buttons are never pressed. Unexpectedly, the tool shows us that this scenario is not allowed. By simulating the requirements and examining a counterexample trace, we find out that the problem lies in requirements (R9) and (R10). In fact, the *until* operator (called *strong until*) used in these requirements forces the button to be pressed eventually, contradicting our intent. To solve the problem, we correct requirements (R9) and (R10) by replacing them with the following ones, which use the *weak until* operator (the *weak until* is satisfied whenever its first argument stays true for the whole trace):

```
(R9-2) forall i in {1:3} : [!service[i] W button[i]]
(R10-2) forall i in {1:3} :
        G(service[i] -> next([!service[i] W button[i]]))
```

Using RAT, we can now check that the new set of requirements is still consistent, that the assertions still hold, and that the possibilities are allowed. In order to allow the desired behaviors, we generalize possibilities (P1) and (P2) as follows:

```
(P3) forall i in {1:3} : F G(!button[i])
```

This possibility states that it is possible that eventually a button is never pressed again. This possibility is allowed by our requirements.

The final set of requirements is shown in Table 5.1. By running further simulations and checking further behaviors, we can finally be reasonably sure that they capture our design intent. Notice that the formalization of the requirements could be carried on, depending on the designer intent. For example, the direction signal could be further refined by distinguishing two different directions, moving_up and moving_down. Moreover, the designer could wish to formalize further properties of the elevator service (e.g., force the elevator to follow an optimal schedule to service a set of incoming requests). ■

5.7.5 Using Model Checking for Property Verification

Model checking techniques can be used to check whether a system satisfies a set of properties that are required for its operation. To this aim, a formal model of the system, in the form of a state transition system (see Section 5.6.5) is built and

Table 5.1   A Set of Requirements for Operating an Elevator

```
(R1)   forall i in {1:3} :
       G (service[i] <-> (floor=i && door=open))
(R2)   forall i in {1:3} : G(button[i] -> F(service[i]))
(R3)   G (door=open -> direction=standing)
(R4)   forall i in {1:3} :
       G(floor=i -> (next(floor) <= i + 1) && next(floor) >= i - 1)
(R5)   forall i in {1:3} :
       G((direction=standing && floor = i) -> (next(floor) = i))
(R6)   forall i in {1:3} :
       G((direction=moving && floor = i) -> (next(floor) != i))
(R7)   forall i in {1:3} : G(button[i] -> [button[i] U service[i]])
(R8)   forall i in {1:3} : G(service[i] -> next(!button[i]))
(R9-2)  forall i in {1:3} : [!service[i] W button[i]]
(R10-2) forall i in {1:3} :
        G(service[i] -> next([!service[i] W button[i]]))
```

then exhaustively analyzed by the model checker to check whether the properties hold. In principle, these techniques can be applied to analyze different kinds of system properties, addressing functional correctness as well as reliability and safety. Interestingly enough, whenever a property does not hold, the model checker can automatically generate a counterexample trace, that is, an execution trace of the system that witnesses the violation of the property. Such kinds of information can be used by design or safety engineers to understand where the problem lies and take appropriate countermeasures.

In the remainder of this section we exemplify the use of model checking for property verification on our case study, namely the Three Mile Island example of Section 1.4. The formal model of the system has been written and verified using the FSAP tool (FSAP, 2009; Bozzano and Villafiorita, 2007) (see Appendix B for more information). FSAP is a toolset for formal verification and safety analysis. It is composed of a graphical user interface and a verification engine, which is implemented on top of the NuSMV model checker. In this section we focus on system behavior under nominal conditions, that is, in the absence of faults. We develop this example further when we discuss formal safety analysis and fault injection in Section 5.8.

Example 5.8   The Three Mile Island case study presented in Section 1.4 describes an example of a highly critical system, namely controlling a nuclear power plant. The set of requirements for its operation are manifold, as they clearly range from functional correctness to ensuring safe operation of the plant. In this section we present a few examples of properties that exemplify the formal verification process. The complete set of properties can be found in the formal model that we provide on the web site http://safety-critical.org. The following properties are expressed in CTL temporal logic (see Section 5.6.6).

The first example we provide is the following property:

```
(TMI1) AG(porv_command = open & block_command = open &
          circuit1.coolant_level <=8 ->
            AG(circuit1.coolant_status != solid))
```

It states that whenever the PORV valve and the associated block valve are commanded open and the level of the coolant in the primary circuit is at least 8, the circuit cannot *go solid*. We verify this property under the following hypotheses:

```
porv_command = open | porv_command = none
block_command = open | block_command = none
```

These hypotheses are imposed as model *invariants*; that is, they must be satisfied in each execution state of the model. In FSAP, hypotheses can be associated with verification tasks; that is, they can be customized for the specific verification problem at hand. In this example, we are requiring that the two valves can never be commanded close in any execution. Using FSAP, we can verify that this property holds.

Another example of property is the following:

```
(TMI2) AG(tank_command = inject & circuit_1.coolant_level > 2 ->
          AG(circuit1.coolant_level >=2))
```

which states that by injecting coolant into the primary circuit using the tanks, the level of the coolant will always remain at least at level 2, provided that initially it is at least at a level greater than 2. This property holds under the following hypothesis:

```
tank_command = inject | tank_command = none
```

That is, we require that the tanks cannot be commanded close.

We can use model checking techniques to verify properties other than functional or safety requirements. Moreover, we can also use model checking techniques to perform guided simulation. For instance, we may want to drive the system into a particular state or verify that a specific state is reachable. As an example, consider the following property:

```
(TMI3) EF(circuit1.coolant_status = solid)
```

stating that there exists a path such that the coolant in the primary circuit can go solid. This property can be verified to hold using FSAP. If we want to generate a trace leading to such a state, we can negate the previous property and check the following:

```
(TMI4) !EF(circuit1.coolant_status = solid)
```

In this case, FSAP returns with a counterexample trace, which is shown in Figure 5.14 (where we reported only the most significant signals and have shortened `porv_command` with `porv_cmd`, `circuit1.coolant_level` with `c1.c_level`, and similarly for the other signals).

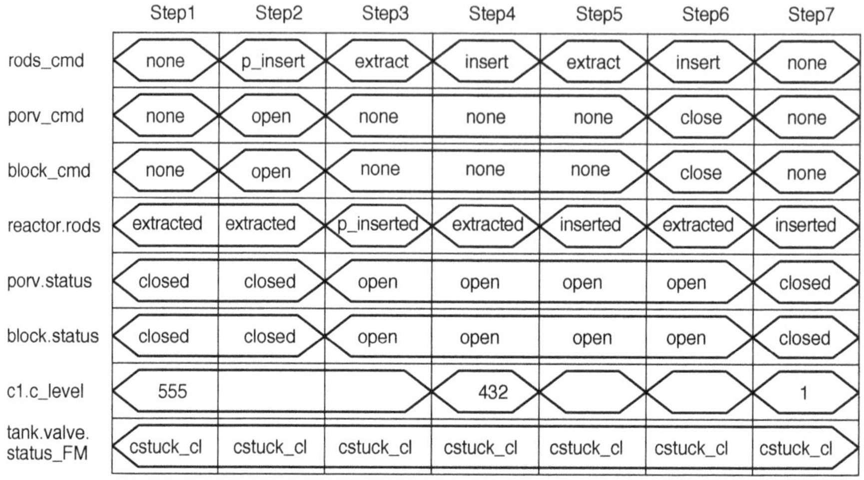

	Step1	Step2	Step3	Step4	Step5	Step6	Step7
rods_cmd	none	p_insert	extract	insert	extract	insert	none
porv_cmd	none	open	none	none	none	close	none
block_cmd	none	open	none	none	none	close	none
reactor.rods	extracted	extracted	p_inserted	extracted	inserted	extracted	inserted
porv.status	closed	closed	open	open	open	open	closed
block.status	closed	closed	open	open	open	open	closed
c1.c_level	555			432			1
tank.valve. status_FM	cstuck_cl	cstuck_cl	cstuck_cl	cstuck_cl	cstuck_cl	cstuck_cl	cstuck_cl

Figure 5.14 A counterexample trace for property (TMI4).

The counterexample trace shows a possible scenario leading to the situation in which the primary circuit goes solid, namely a scenario in which the block valve is commanded to close and the tank is commanded to inject liquid into the primary circuit. As a consequence, the level of the coolant in the primary circuit eventually reaches level 10, causing it to go solid.

As the last example, we verify the following property:

```
(TMI5) AG((pump1a.status = broken & pump1b.status = broken) ->
          AF(reactor.core_status = 5))
```

under the hypothesis that the control rods are not operated:

```
rods_command = none
```

In other words, we are stating that if the pumps of the primary circuit break and we do not insert the rods, eventually the reactor core will reach the critical level of 5. This property is found to be false by FSAP. In fact, the stable plant hypothesis ensures that an increase in the temperature of the primary circuit eventually slows down the nuclear reaction, preventing the core from reaching the critical level. A counterexample to this property is given by an infinite path, in which eventually the reactor core is always below the critical level. The corresponding FSAP trace represents this behavior in a finite manner with a loop, in which the reactor core stabilizes to level 3. ∎

5.8 Formal Safety Analysis

In this section we discuss the application of formal methods to safety assessment. As discussed in Chapter 4, safety assessment is an integral and essential part of the development life cycle of critical systems. Although the use of formal methods as an aid to support verification and validation activities in the development process has attracted increasing interest over the years, the use of formal methods to produce artifacts and support activities that are specific of safety assessment, such as techniques for hazard analysis, is relatively new. Seminal work in this field has been carried out within the ESACS (2009) (Enhanced Safety Assessment for Complex Systems) and ISAAC (2009) (Improvement of Safety Activities on Aeronautical Complex systems) projects, two European-Union-sponsored projects involving various research centers and industries from the avionics sector. The key ingredient of ESACS and ISAAC is a comprehensive methodology (called the ESACS methodology), supported by formal methods, to assist both system development and safety analysis, and tool-supported verification and validation. More details on these projects can be found in Bozzano et al. (2003) and Åkerlund et al. (2006). A related approach is presented in Joshi and Heimdahl (2005) and Joshi et al. (2005), where the authors propose to integrate formal safety analysis into the traditional "V" safety assessment process.

The ESACS methodology (Bozzano et al., 2003; Bozzano and Villafiorita, 2007) is based on model-based formal safety analysis, that is, on the use of formal methods for both system modeling and verification. Model-based safety analysis aims to reduce the effort involved in safety assessment and increase the quality of the results by focusing the effort on building formal models of the system, rather than carrying out the analyses. Formal safety analysis is carried out on models that take into account system behavior in the presence of faults. In particular, models are formalized as state transition systems (see Section 5.6.5) and model analysis uses symbolic model checking techniques. Furthermore, shared formal notations are used as the common language between the design and safety analysis stages.

The ESACS methodology is summarized in Figure 5.15. It is based on a number of steps. The starting point is a formal model, which can be written either by the design engineer or by the safety engineer (Bozzano and Villafiorita, 2007). In the remainder of this section we focus on the former scenario; that is, the one in which the model is written by the design engineer. This model is called **nominal system model**, because it includes only the nominal behavior of the system; that is, faults are not taken into account. This model can be used by the design engineer, for instance to verify functional requirements, and it is then handed over to the safety engineer to perform safety assessment. As advocated in Bozzano et al. (2003), it is important to have a complete decoupling between the nominal system model and the fault model. For this reason, the ESACS methodology is based on the concept of **fault injection**.

The mechanism of fault injection is responsible for extending a formal model, corresponding to the nominal model of the system under analysis, in order to take into account the presence of faults. The fault injection step is realized automatically; it takes as input a nominal model and a specification of the failure modes to be injected, and produces a new formal model. The model resulting after fault injection is called **extended model**, and the extension process is called **model extension** (Bozzano and Villafiorita, 2007). Typically, the failure modes to be injected can be retrieved

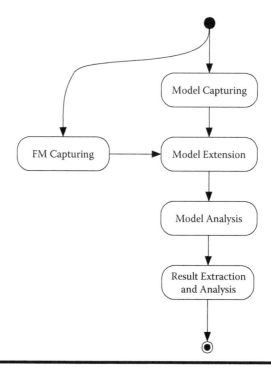

Figure 5.15   The ESACS methodology.

from a library that contains a set of predefined failure modes, or they can be defined manually. The extended model can be used by the safety engineers to define safety requirements and perform all sorts of analyses related to hazard analysis and safety assessment.

In the remainder of this section we exemplify formal safety analysis techniques using the FSAP tool (Bozzano and Villafiorita, 2007; FSAP, 2009). FSAP implements the ESACS methodology (for other tools implementing the same methodology, see, for example, Bozzano et al. (2003), Bieber et al. (2002), Abdulla et al. (2004), Deneux and Åkerlund (2004), and Peikenkamp et al. (2004)). The remainder of this section is organized as follows. We first discuss the concept of fault injection in Section 5.8.1 and the related concept of model extension in Section 5.8.2. We then discuss the use of formal safety analysis techniques for hazard analysis, in particular FTA in Section 5.8.4 and FMEA in Section 5.8.5.

5.8.1 Fault Injection

The mechanism of fault injection is responsible for the model extension step. In Figure 5.16 we illustrate this mechanism pictorially. On the left-hand side, we assume a model containing a functional block called "A." The block has two inputs signals

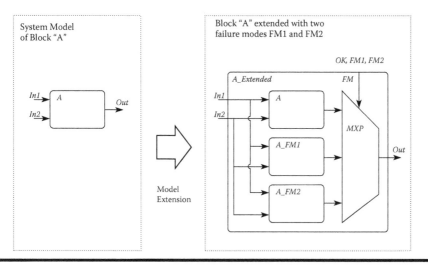

Figure 5.16   Model extension.

In1 and *In2* and an output signal *Out*. We want to extend "A" by injecting two faults FM1 and FM2.

On the right-hand side, the extended model is illustrated. It consists of a functional block called "A_Extended" that contains the original block "A" and two additional blocks "A_FM1" and "A_FM2," which formalize, respectively, the effect of the two faults FM1 and FM2 on the functional block "A." For instance, if FM1 is a *stuck_at zero* fault, the output of the block A_FM1 will produce the value *zero* independently of its inputs. We note that both block "A" and the new blocks "A_FM1" and "A_FM2" take the same inputs as the original block "A" on the right-hand side. The *Out* signal in the extended model is obtained as output of a multiplexer ("MXP"), which takes as inputs the outputs of the three blocks "A," "A_FM1," and "A_FM2." The output of the multiplexer is controlled by the *FM* signal, called the *failure mode variable*, which formalizes the fault status of the "A" block and can assume a value in the set {*OK,FM1,FM2*}. If *FM* has value *OK*, then no fault has occurred and the output of block "A" is copied to the output *Out*. If *FM* has value *FM1*, then it is intended that fault FM1 has occurred and, as a consequence, the output of block "A_FM1" is copied to the output *Out*. Similarly for the *FM2* case.

In general, if a model contains several blocks, fault injection can be applied to each block independently. For each block, a different failure mode variable is introduced.

5.8.2 Fault Models and Model Extension

The fault injection mechanism described in the previous section is independent of the specific fault model of the fault being injected. The specification of the fault model is hidden in the blocks implementing the faulty behavior (blocks "A_FM1" and "A_FM2" in Figure 5.16) and in the specification of the behavior of the failure mode variable (*FM* in Figure 5.16). The latter encodes, for instance, whether the fault is permanent or transient.

We illustrate the definition of a fault model and the model extension mechanism using an example written in the NuSMV language and automatically extended with the FSAP platform. The FSAP platform provides a predefined library of failure modes, called the Generic Failure Mode Library, comprising, among others, the following failure mode types: *inverted* (for Boolean variables), *stuck_at* ⟨*value*⟩, *cstuck_at* ⟨*value*⟩, *random* (nondeterministic output), and *ramp_down* (for integer-valued variables). The *cstuck_at* failure mode, where "*c*" stands for *continuity*, is a variant of *stuck_at*, which imposes additional constraints on the occurrence of the failure mode, namely that a variable can fail stuck at value *v* only when its nominal value is *v* (e.g., this allows us to enforce that a valve can fail stuck open only when it is open, not when it is closed). In FSAP, faults can be declared as either permanent or sporadic; in the latter case, a fault is allowed to show up itself only transiently, or to disappear as a consequence of a repair.

Example 5.9 Table 5.2 presents a simple example of a functional block written in NuSMV. The block, called BIT, takes one input Boolean signal In and generates one Boolean output signal Out, which is identical to In. The fault to be injected is a random fault of type *inverted*. Hence, after fault injection, the Out signal of the extended model will be inverted with respect to the In signal. We assume the fault to be permanent.

The extended model generated by FSAP and encoded in NuSMV is illustrated in Table 5.3. In the extended model (lines 3–7) a new variable Out_nominal is declared to behave the same as the original Out variable; it represents the behavior of the output under nominal conditions. The behavior of the faulty output Out_inverted (line 9) is defined to be inverted with respect to the nominal one. The failure mode variable Out_FailureMode is declared in lines 16 and 17; it can assume the values no_failure or inverted (the former represents nominal behavior, whereas the latter represents the behavior after occurrence of the fault). The new Out variable (lines 11–14) is defined to behave as the nominal variable Out_nominal when the failure mode variable has value no_failure,

Table 5.2   A Simple Functional Block in NuSMV

```
1      MODULE BIT(In)
2
3      VAR
4        Out: boolean;
5
6      ASSIGN
7        Out := In;
```

Table 5.3 The BIT Block Extended with an Inverted Failure Mode

```
1    MODULE BIT(In)
2
3    VAR
4      Out_nominal: boolean;
5
6    ASSIGN
7      Out_nominal := In;
8
9    DEFINE Out_inverted := ! Out_nominal;
10
11   DEFINE Out := case
12     Out_FailureMode = no_failure : Out_nominal;
13     Out_FailureMode = inverted : Out_inverted;
14   esac;
15
16   VAR
17     Out_FailureMode : {inverted, no_failure};
18
19   ASSIGN
20     next(Out_FailureMode) := case
21       Out_FailureMode = no_failure : {inverted, no_failure};
22       1: Out_FailureMode;
23   esac;
```

whereas it behaves as the faulty output variable `Out_inverted` when the failure mode variable has value `inverted`. Finally, lines 19 through 23 encode the fault model. The fault can randomly occur (line 21). Moreover, in this example the fault is permanent (line 22); hence the failure mode variable keeps the value `inverted` forever, once the fault has occurred. ∎

Both the nominal model and the extended model in the previous example are NuSMV models. In general, they can both be represented as Kripke structures (compare Section 5.6.5). Formally, we can describe the process of model extension as follows. Let $\mathcal{M} = \langle \mathcal{S}, \mathcal{I}, \mathcal{R}, \mathcal{L} \rangle$ be a Kripke structure representing the nominal system model. Model extension takes as input a specification of the faults to be added, and automatically generates the Kripke structure corresponding to the extended system model. A fault is defined by the proposition $p \in \mathcal{P}$ to which it must be attached, and by its type, which specifies the faulty behavior of p in the extended system (e.g., p has type *inverted* in the previous example). Model extension introduces a new proposition p^{FM}, the failure mode variable, modeling the possible occurrence of the fault, and two further propositions p^{Failed} and p^{Ext}, with the following intuitive meaning. The proposition p^{Failed} models the behavior of p when a fault has occurred. For instance, the following condition (where $\mathcal{S}'$ is the set of states of the extended system model) is used to define an *inverted* failure mode (that is, p^{Failed} holds if and only if p does not hold):

$$\forall s \in \mathcal{S}' \quad (s \models p^{Failed} \iff s \not\models p) \tag{5.11}$$

The proposition p^{Ext} models the extended behavior of p; that is, it behaves as the original p when no fault is active, whereas it behaves as p^{Failed} in the presence of a fault:

$$\forall s \in \mathcal{S}' \quad s \not\models p^{FM} \rightarrow (s \models p^{Ext} \iff s \models p) \tag{5.12}$$

$$\forall s \in \mathcal{S}' \quad s \models p^{FM} \rightarrow (s \models p^{Ext} \iff s \models p^{Failed}) \tag{5.13}$$

The extended system model $\mathcal{M}^{Ext} = \langle \mathcal{S}', \mathcal{I}', \mathcal{R}', \mathcal{L}' \rangle$ can be defined in terms of the nominal system model by adding the new propositions, modifying the definition of the (initial) states and of the transition relation, and imposing the Conditions (5.11) (for an *inverted* failure mode), (5.12), and (5.13). We omit the details for the sake of simplicity. Finally, model extension with respect to a *set* of propositions can be defined in a straightforward manner by iterating model extension over single propositions.

In the remainder of this chapter we assume that a Kripke structure $\mathcal{M} = \langle \mathcal{S}, \mathcal{I}, \mathcal{R}, \mathcal{L} \rangle$ over a set of propositions $\mathcal{P}$, representing an extended model, is given. We also assume that the set of failure mode variables of the model is given by the set $\mathcal{F} \subseteq \mathcal{P}$, and we use the term **fault configuration** to denote a subset of failure mode variables.

5.8.3 Property Verification

In this section, drawing a parallel with Section 5.7.5, we illustrate how model checking techniques can be used to verify system properties in the presence of faults. Using FSAP, this is accomplished by defining appropriate fault models, performing fault injection, and carrying out the analyses on the extended model.

We illustrate property verification, using again the Three Mile Island example of Section 1.4. In the remainder of this chapter, we assume the nominal model to be enriched with the definition of the following failure modes:

■ A *stuck_at broken* failure mode for each of the pumps (the two pumps for the primary circuit (called *pump1a* and *pump1b*), the two pumps for the secondary circuit (*pump2a* and *pump2b*), and the tank pump). We notice that in the nominal model it is already possible for the pumps to break as a result of tearing due to the presence of air bubbles in the coolant. The newly added failure mode must be considered a primitive failure mode; that is, it can occur as the result of a primitive failure event that is independent of the working conditions of the pumps.

■ Two *cstuck_at open* and *cstuck_at close* failure modes for each of the two valves (the PORV valve and the block valve). Again, this must be considered a primitive failure mode causing a valve to get stuck either open or closed. Notice that by using the *cstuck* fault model, we require that an open valve can only fail stuck open and, similarly, a closed valve can only fail stuck closed.

All the faults are declared as being permanent.

Example 5.10 We verify again the property (TMI1) of Section 5.7.5 (which we rename as TMI1'):

```
(TMI1') AG(porv_command = open & block_command = open &
           circuit1.coolant_level <=8 ->
             AG(circuit1.coolant_status != solid))
```

under the usual hypotheses:

```
porv_command = open | porv_command = none
block_command = open | block_command = none
```

By running FSAP, we can check that the property still holds when verified in the extended model. That is, the primary circuit cannot *go solid*, under the given hypotheses, even in presence of faults of the pumps or the valves.

We get a a different outcome if we check property (TMI2) (renamed TMI2'):

```
(TMI2') AG(tank_command = inject & circuit_1.coolant_level > 2 ->
             AG(circuit1.coolant_level >=2))
```

under the usual hypothesis:

```
tank_command = inject | tank_command = none
```

This time, the property does not hold, and the model checker returns with the counterexample shown in Figure 5.17 (as usually, we have included only the most significant signals and shortened some of them). Evidently, this trace shows that property (TMI2') can be violated as a result of a failure of the tank valve. If the valve fails stuck closed, then the liquid contained in the tank cannot be injected in the primary circuit, eventually causing the coolant level to decrease to under level 2.

We can verify that breaking of the tank pump is not the only fault that can cause violation of property (TMI2'). In fact, if we add the following hypothesis:

```
tank.valve.status_FailureMode = no_failure
```

and we run FSAP, we discover that property (TMI2') can also be violated due to breaking of the tank pump. However, if we rule out this failure by adding the following additional hypothesis:

```
tank.pump.status_FailureMode = no_failure
```

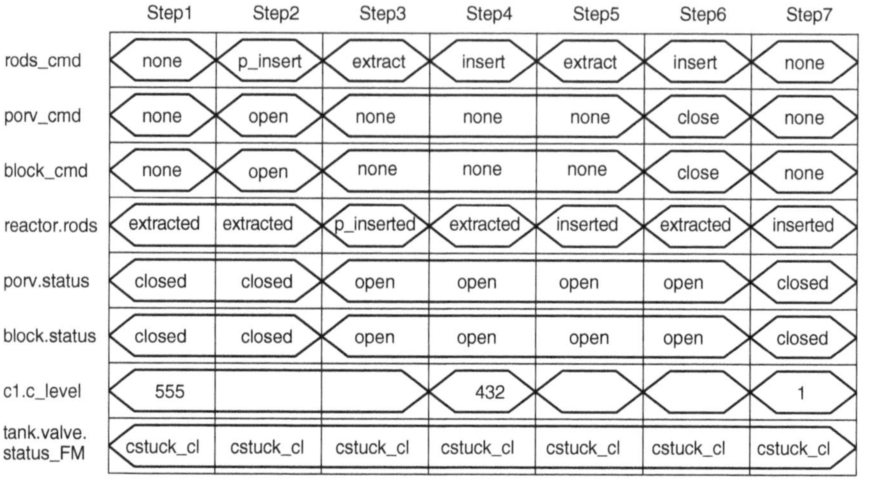

	Step1	Step2	Step3	Step4	Step5	Step6	Step7
rods_cmd	none	p_insert	extract	insert	extract	insert	none
porv_cmd	none	open	none	none	none	close	none
block_cmd	none	open	none	none	none	close	none
reactor.rods	extracted	extracted	p_inserted	extracted	inserted	extracted	inserted
porv.status	closed	closed	open	open	open	open	closed
block.status	closed	closed	open	open	open	open	closed
c1.c_level	555			432			1
tank.valve.status_FM	cstuck_cl	cstuck_cl	cstuck_cl	cstuck_cl	cstuck_cl	cstuck_cl	cstuck_cl

Figure 5.17   A counterexample trace for property (TMI2′).

that is, we assume that both the pump and the valve cannot fail, then property (TMI2') can be verified to hold.   ∎

5.8.4 Fault Tree Generation

As explained in Section 3.2.1, Fault Tree Analysis (Vesely et al., 1981, 2002) is an example of deductive analysis, which, given a *top-level event* (TLE), that is, the specification of an undesired condition, builds all possible chains of one or more basic faults that contribute to the occurrence of the event, and pictorially represents them in a so-called *fault tree*. In its simpler form (Bozzano and Villafiorita, 2007), a fault tree can be represented as the collection of its *minimal cut sets*. In this form, a fault tree has a two-layer logical structure, consisting of the top-level disjunction of the minimal cut sets, each cut set being the conjunction of the corresponding basic faults. In this section we describe how such fault trees can be generated using formal safety analysis techniques, and in particular we exemplify this process using the FSAP tool. In general, logical structures with multiple layers can be used, for instance, based on the hierarchy of the system model (Banach and Bozzano, 2006).

The framework described here is completely general; it encompasses both the case of *permanent* faults (*once failed, always failed*) and the case of *sporadic* or *transient* ones—that is, when faults are allowed to occur sporadically (e.g., a sensor showing an invalid reading for a limited period of time), possibly repeatedly over time, or when repairing is possible.

We first give the following definitions of (minimal) cut set.

Definition 5.3 (Cut set) Let $\mathcal{M} = \langle \mathcal{S}, \mathcal{I}, \mathcal{R}, \mathcal{L} \rangle$ be a Kripke structure with a set of failure mode variables $\mathcal{F} \subseteq \mathcal{P}$, let $FC \subseteq \mathcal{F}$ be a fault configuration, and $TLE \in \mathcal{P}$. We say that FC is a cut set of TLE, written $cs(FC, TLE)$ if there exists a trace $s_0, s_1, \ldots, s_k$ for $\mathcal{M}$ such that: *i*) $s_k \models TLE$; *ii*) $\forall f \in \mathcal{F}$ $f \in FC \iff \exists i \in \{0, \ldots, k\}$ $(s_i \models f)$.

Intuitively, a cut set corresponds to the set of failure mode variables that are active at some point along a trace witnessing the occurrence of the top-level event. Typically, one is interested in isolating the minimal cut sets, that is, the fault configurations that are minimal in terms of fault variables. **Minimal cut sets** are formally defined as follows.

Definition 5.4 (Minimal Cut Sets) Let $\mathcal{M} = \langle \mathcal{S}, \mathcal{I}, \mathcal{R}, \mathcal{L} \rangle$ be a Kripke structure with a set of failure mode variables $\mathcal{F} \subseteq \mathcal{P}$, let $F = 2^{\mathcal{F}}$ be the set of all fault configurations, and $TLE \in \mathcal{P}$. We define the set of cut sets and minimal cut sets of TLE as follows:

$$CS(TLE) = \{FC \in F \mid cs(FC, TLE)\} \tag{5.14}$$

$$MCS(TLE) = \{cs \in CS(TLE) \mid \forall cs' \in CS(TLE) \ (cs' \subseteq cs \to cs' = cs)\} \tag{5.15}$$

The notion of minimal cut set can be extended to the more general notion of **prime implicant** (see Coudert and Madre (1993)), which is based on a different definition of minimality, involving both the activation and the absence of faults. Based on the previous definitions, Fault Tree Analysis can be described as the activity that, given a TLE, involves the computation of all the minimal cut sets (or prime implicants) and their arrangement in the form of a tree. Both the computation of minimal cut sets and of prime implicants are supported by the FSAP platform. Typically, each cut set is also associated with a trace witnessing the occurrence of the TLE.

In the remainder of this section we discuss the symbolic implementation of an algorithm for the generation of fault trees and its implementation in the FSAP tool. In particular, we present the basic algorithm for *forward fault tree analysis*, that is, a routine for generating fault trees using forward reachability. An alternative algorithm, based on backward reachability, and a discussion about some important optimizations that significantly improve the performance of these routines are described in Bozzano et al. (2007). We focus on the computation of the cut sets (standard model checking techniques can be used to generate the corresponding traces).

In the following we use the following notations. First, $\underline{f}$ denotes the vector of failure mode variables. Furthermore, given two vectors $\underline{v} = v_1, \ldots, v_k$ and $\underline{w} = w_1, \ldots, w_k$, the notation $\underline{v} = \underline{w}$ stands for $\bigwedge_{i=1}^{k}(v_i = w_i)$. Finally, we use $ITE(p, q, r)$ (*if then else*) to denote $((p \to q) \land (\neg p \to r))$.

The forward algorithm starts from the set of initial states of the system and accumulates, at each iteration, the forward image (compare Section 5.7.3). To take into account sporadic faults (compare Definition 5.3, condition *ii*), at each iteration we need to "remember" if a fault has been activated. To this aim, for each failure mode variable $f_i \in \mathcal{F}$, we introduce an additional variable o_i (*once* f_i), which is true if and only if f_i has been true at some point in the past. This construction is traditionally referred to as *history variable* and is formalized by the transition relation $\mathcal{R}^o$ given by the following condition:

$$\bigwedge_{f_i \in \mathcal{F}} ITE(o_i, o'_i, o'_i \leftrightarrow f'_i) \qquad (5.16)$$

Let *Extend*$(\mathcal{M}, \mathcal{R}^o)$ be the Kripke structure obtained from $\mathcal{M}$ by replacing the transition relation $\mathcal{R}$ with the synchronous product between $\mathcal{R}$ and $\mathcal{R}^o$, in symbols $\mathcal{R}(\underline{x}, \underline{x}') \wedge \mathcal{R}^o(\underline{x}, \underline{x}')$ and modifying the labeling function $\mathcal{L}$ accordingly.

The pseudo-code of the algorithm is described in Table 5.4. The inputs are $\mathcal{M}$ and *Tle* (the set of states satisfying the TLE). A variable *Reach* is used to accumulate the reachable states, and a variable *Front* to keep the *frontier*, that is, the newly generated states (at each step, the image operator needs to be applied only to the latter set). Both variables are initialized with the initial states, and the history variables with the same value as the corresponding failure mode variables. The core of the algorithm (lines 4–8) computes the set of reachable states by applying the image operator to the frontier until a fixpoint is reached (i.e, the frontier is the empty set). The resulting set is intersected (line 9) with *Tle*, and projected over the history variables. Finally, the minimal cut sets are computed (line 10), and the result is mapped back from the history variables to the corresponding failure mode variables (line 11).

Table 5.4 A Symbolic Algorithm for Fault Tree Generation

function FTA-Forward $(\mathcal{M}, Tle)$
1      $\mathcal{M} := Extend(\mathcal{M}, \mathcal{R}^o)$;
2      $Reach := \mathcal{I} \cap (\underline{o} = \underline{f})$;
3      $Front := \mathcal{I} \cap (\underline{o} = \underline{f})$;
4   **while** $(Front \neq \emptyset)$ **do**
5         $temp := Reach$;
6         $Reach := Reach \cup fwd\_img(\mathcal{M}, Front)$;
7         $Front := Reach \backslash temp$;
8   **end while**;
9      $CS := Project(\underline{o}, Reach \cap Tle)$;
10     $MCS := Minimize(CS)$;
11   **return** $Map_{\underline{o} \rightarrow \underline{f}}(MCS)$;

Note that all the primitives used in algorithm can be realized using BDD data structures, as explained in Section 5.7.3. For instance, set difference (line 7) can be defined as $Reach(\underline{x}) \wedge \neg temp(\underline{x})$, and projection (line 9) as $\exists \underline{y}.(Reach(\underline{x}) \wedge Tle(\underline{x}))$, where $\underline{y}$ is the set of variables in $\underline{x}$ and not in $\underline{o}$; the minimization routine (line 10) can be realized with standard BDD operations (as described in Coudert and Madre (1993) and Rauzy (1993)); finally, the mapping function (line 11) can be defined as $Map_{\underline{o} \rightarrow \underline{f}} \phi(\underline{o}) = \exists \underline{o}.(\phi(\underline{o}) \wedge (\underline{o} = \underline{f}))$.

We now exemplify fault tree generation with the FSAP tool on the Three Mile Island example of Section 1.4.

Example 5.11   Using FSAP, we want to find out which combinations of primitive faults can cause the reactor core to melt; that is, we define the following TLE:

```
(TLE1) reactor.melted
```

In general, if we do not constrain the way the plant is operated, it is always possible to achieve melting of the reactor core in nominal conditions. In other words, it is possible to have it melted by operating the plant in an intentionally wrong way (to this aim, it is sufficient to let the temperature of the primary circuit increase to a dangerous level by stopping all the pumps, and then drain the primary circuit). We therefore want to carry out FTA under proper hypotheses for the way the plant is controlled. In particular, we specify the following set of hypotheses:

```
command1 != stop & command2 != stop
porv_command = open -> circuit1.temp_output_low >= 4
circuit1.coolant_level <= 3 -> porv_command = close
```

The first hypothesis requires that the pumps of the primary and the secondary systems cannot be commanded to stop. The remaining two hypotheses constrain the conditions under which the PORV valve should be commanded to open or close. In particular, it is required that the PORV valve *can* only be opened when the temperature of the primary circuit increases above level 4; moreover, whenever the coolant level of the primary circuit decreases under level 3, the PORV valve *must* be commanded to close.

Using the FSAP tool, we can automatically generate the fault tree shown in Figure 5.18.

The fault tree shows that melting of the reactor core can follow as a result of two distinct combinations of faults (minimal cut sets):

■ The first combination requires the two parallel pumps connected to the primary circuit to fail broken, and the PORV valve to fail stuck opened.
■ The second combination requires the two parallel pumps connected to the secondary circuit to fail broken, and the PORV valve to fail stuck opened.

The second combination of faults actually caused partial melting of the reactor core at Three Mile Island in 1979 (see Section 1.4). To better understand the sequence of

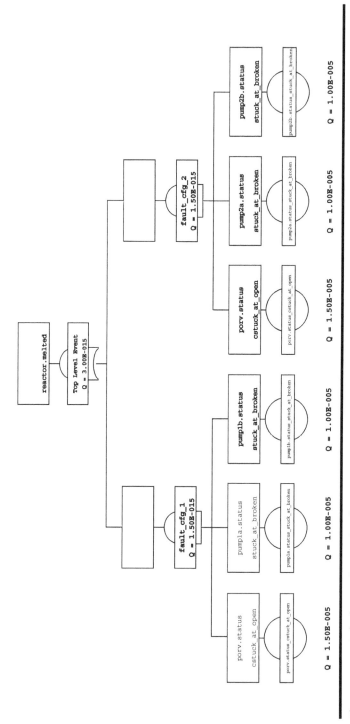

Figure 5.18 A fault tree for top level event (TLE1).

events that can lead to such an accident, we can examine the witness traces generated by FSAP for each of the minimal cut sets. We focus our attention on the second cut set, as the first one is very similar. The corresponding trace is shown in Figure 5.19. We notice that at step 2, both pumps in the secondary circuit, `pump2a` and `pump2b`, fail broken. As a consequence, the temperature of the primary circuit, `circuit1.temp_output_low`, which cannot be cooled down anymore from the coolant of the secondary circuit, begins to increase, eventually reaching level 4 at step 8. The PORV valve is commanded to open at step 8, and it is in fact opened at step 9, causing the coolant level `circuit1.coolant_level` in the primary circuit to decrease, as expected. At step 11, the coolant level reaches level 3, and the PORV valve is commanded to close, as required by the hypotheses on the controller. However, at step 11, the PORV valve fails stuck opened, which prevents it from closing. Eventually, at step 13, the coolant level reaches level 1, while the temperature of the primary circuit is 4, causing the reactor to melt.   ∎

We conclude this section by remarking that, in dynamic systems, different fault models (e.g., sporadic versus permanent) may have different impacts on the results. Moreover, the duration and temporal relationships may be important; for example, fault f_1 may be required to persist for a given number of steps, or to occur before another fault f_2. A proposal to enrich the notion of minimal cut set along these lines can be found in Åkerlund et al. (2006). At the moment, there are two ways in which FSAP can deal with dynamic information that is not visible within standard fault trees. First, for each cut set, it can generate a counterexample trace that witnesses the violation of the TLE. Second, it supports the so-called **failure ordering analysis**, whose purpose is to investigate the ordering constraints between basic events within a cut set. More details can be found in Bozzano and Villafiorita (2003).

5.8.5 FMEA Table Generation

Failure Mode and Effects Analysis is a classical technique for hazard analysis. As explained in Section 3.2.2, it is an example of an inductive technique; that is, it starts by identifying a set of failure modes of the system and, using forward reasoning, assesses their impact on a set of events. Inputs to FMEA are therefore a set of failure modes and a set of events to be analyzed. Typically, FMEA considers only single faults, although fault configurations involving several faults can be considered in particular cases. The set of fault configurations can be provided implicitly; for instance, it is possible to analyze the effect of all the fault configurations that are combinations of at most two faults taken from the set of input failure modes. The results of the analysis are summarized in a so-called FMEA table, which links the input fault configurations with their effect on the input events.

In this section we use the FSAP tool to exemplify the generation of FMEA tables in formal safety analysis. We use the same framework described in Section 5.8.4. In particular, we assume that a set of fault configurations $FC_1, \ldots, FC_n$ and a

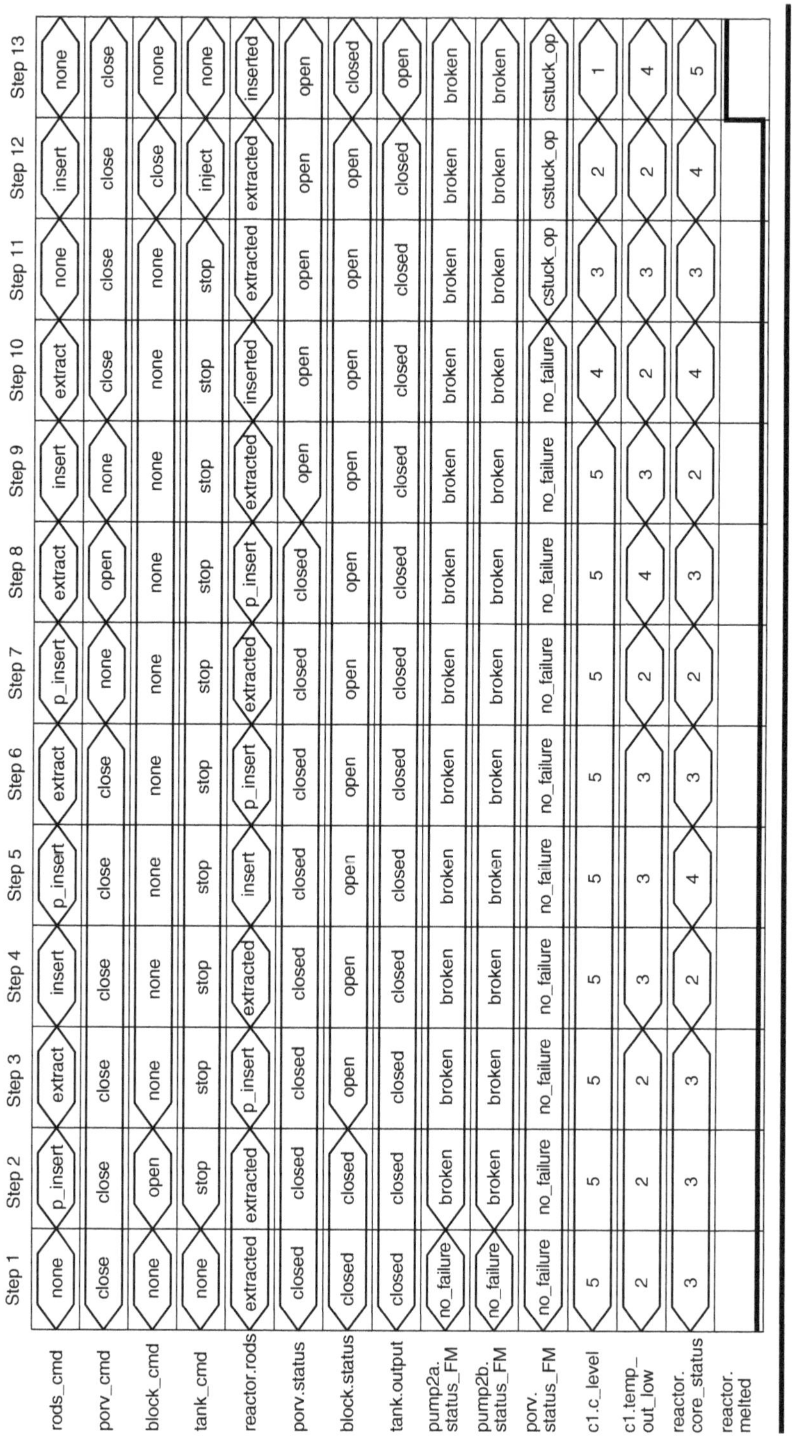

	Step 1	Step 2	Step 3	Step 4	Step 5	Step 6	Step 7	Step 8	Step 9	Step 10	Step 11	Step 12	Step 13
rods_cmd	none	p_insert	extract	insert	p_insert	extract	p_insert	extract	insert	extract	none	insert	none
porv_cmd	close	close	close	close	close	close	none	open	none	close	close	close	close
block_cmd	none	open	none	none	none	none	none	none	none	none	none	close	none
tank_cmd	none	stop	stop	stop	stop	stop	stop	stop	stop	stop	stop	inject	none
reactor.rods	extracted	extracted	p_insert	extracted	insert	p_insert	extracted	p_insert	extracted	inserted	extracted	extracted	inserted
porv.status	closed	closed	open	open	closed	closed	closed	closed	open	open	open	open	open
block.status	closed	closed	open	open	open	open	open	open	open	open	open	open	closed
tank.output	closed	closed	closed	closed	closed	closed	closed	closed	closed	closed	closed	closed	open
pump2a. status_FM	no_failure	broken	broken	broken	broken	broken	broken	broken	broken	broken	broken	broken	broken
pump2b. status_FM	no_failure	broken	broken	broken	broken	broken	broken	broken	broken	broken	broken	broken	broken
porv. status_FM	no_failure	no_failure	no_failure	no_failure	no_failure	no_failure	no_failure	no_failure	no_failure	no_failure	cstuck_op	cstuck_op	cstuck_op
c1.c_level	5	5	5	5	5	5	5	5	5	4	3	2	1
c1.temp_ out_low	2	2	2	3	3	3	2	4	3	2	3	2	4
reactor. core_status	3	3	3	2	4	3	2	3	2	4	3	4	5
reactor. melted													

Figure 5.19   A witness trace for the second cut set of the fault tree in Figure 5.18.

set of events $E_1, \ldots, E_m \in \mathcal{P}$ is given. Again, the framework is general enough to accommodate both permanent and sporadic faults. The set of entries of the FMEA table is determined by Definition 5.3; that is, a pair (FC_j, E_l) is an entry of the FMEA table if FC_j is a cut set for E_l. Formally, we have the following definition.

Definition 5.5 (FMEA table) Let $\mathcal{M} = \langle \mathcal{S}, \mathcal{I}, \mathcal{R}, \mathcal{L} \rangle$ be a Kripke structure with a set of failure mode variables $\mathcal{F} \subseteq \mathcal{P}$, let $FC_j \subseteq \mathcal{F}$ for $j = 1, \ldots, n$ be a set of fault configurations, and $E_l \in \mathcal{P}$ for $l = 1, \ldots, m$. An FMEA table for $\mathcal{M}$ is the set of pairs $\{(FC_j, E_l) \mid cs(FC_j, E_l)\}$.

We exemplify the generation of FMEA tables with the FSAP tool on the Three Mile Island example of Section 1.4.

Example 5.12 Using FSAP, we want to assess the impact of faults on the following set of events:

```
reactor.melted
circuit1.temp_output_low = 5
circuit1.coolant_level = 1
```

The set of events include the melting of the reactor core, the temperature of the primary circuit raising up to level 5, and the coolant level of the primary circuit dropping to level 1. Similar to Section 5.8.4, we constrain the model with the following additional hypotheses on the way the plant is controlled:

```
command1 != stop & command2 != stop
porv_command = open -> circuit1.temp_output_low >= 4
circuit1.coolant_level <= 3 -> porv_command = close
```

By default, FSAP generates an FMEA table considering only single faults. By setting a parameter, it is possible to analyze fault configurations of an arbitrary order n (that is, fault configurations including at most n primitive faults). In this example, a default run of FSAP produces an empty FMEA table. In other words, under the given hypotheses, no single faults can cause any of the events to occur. If we increase the order to two, we obtain the FMEA table shown in Table 5.5 (where we have shortened pump1a.status_FailureMode = stuck_at_broken with pump1a_broken, and similarly for the other failure modes).

Table 5.5 An FMEA Table with Order 2

Id.Nr.	Failure Mode	Failure effects
1	(pump1a_broken & pump1b_broken)	circuit1.temp_output_low = 5
2	(pump2a_broken & pump2b_broken)	circuit1.temp_output_low = 5

Table 5.6 An FMEA Table with Order 3

Id.Nr.	Failure Mode	Failure effects
1	(porv_cstuck_at_open & pump1a_broken & pump1b_broken)	reactor.melted
2	(porv_cstuck_at_open & pump1a_broken & pump1b_broken)	circuit1.coolant_level = 1
3	(porv_cstuck_at_open & pump2a_broken & pump2b_broken)	reactor.melted
4	(porv_cstuck_at_open & pump2a_broken & pump2b_broken)	circuit1.coolant_level = 1

Each entry in the FMEA table is characterized by an identification number (first column); the failure mode, describing the fault configuration (second column); and the failure effects, listing the set of events impacted by the corresponding fault configuration (third column). In this example, we can observe that the only event that may occur as a consequence of double faults is the temperature of the primary circuit raising up to level 5, and the corresponding cause is breaking of both pumps in either the primary or the secondary circuit.

Finally, if we increase the order to three, we obtain the FMEA table shown in Table 5.6 (where we have shortened `porv.status_FailureMode = cstuck_at_open` into `porv_cstuck_at_open` and, for conciseness, we have omitted the event `circuit1.temp_output_low = 5` from the analysis). Both the events under analysis are caused by the same fault configurations, and the results are in accordance with the results of Fault Tree Analysis (compare Example 5.11). ∎

5.9 Industrial Applications of Formal Methods

We conclude this chapter by looking at some examples of the use of formal method techniques in industry. They range from examples that pioneered the application of formal methods, to more recent examples that illustrate the current state of the art. Although the list is far from being complete, these examples illustrate both the diversity of the industrial fields in which formal methods have been applied and the variety of the formalisms that have been used. We have included examples that are representatives of each of the different classes of formal languages presented in Section 5.6 (see also Rushby (1993), Clarke and Wing (1996), and Bowen and Stavridou (1992)).

5.9.1 IBM's Customer Information Control System (CICS)

A paradigmatic example of the application of formal methods, often cited as one of the most successful ones, was the formal verification of IBM's Customer Information Control System and online transaction processing system (Houston and King, 1991). The project was conducted in the 1980s as a collaboration between Oxford University and IBM Hursley Laboratories. The overall system consisted of more than 750,000 lines of code, about one third of them modified in the latest release of the software. A portion of the code was produced from Z specifications, or partially specified in Z, and the resulting specifications were verified using a rigorous approach (according to the classification of Section 5.3). Some tools that were made available during the projects, such as tools for parsing and typechecking, were used to assist specifiers and code inspectors. Overall, more than 2,000 pages of formal specifications were developed. IBM reported an improvement in the quality of the product, measured as a reduction in the number of faults found by the customers in the part of the code that was formally verified (2:5 ratio with respect to parts of the code that were not subject to formal verification), and an earlier detection of the faults during development. Furthermore, it was estimated that the use of formal methods led to a reduction in the development costs of the new release of the software of about 9%.

5.9.2 The Central Control Function Display Information System (CDIS)

The Central Control Function Display Information System was delivered from Praxis to the U.K. Civil Aviation Authority in 1992, as part of a new air traffic management system for London's airspace (Hall, 1996). The overall system consists of a distributed network of about a hundred computers, implemented using a fault-tolerant architecture, together with a central processing unit and several workstations. Formal methods were used, in different forms and levels of application, throughout the development process, in combination with more traditional software engineering and quality control techniques. The requirements analysis phase was assisted by formal descriptions using structured notations. System specification included a formal description of the data and operations carried out by the system, written in VDM. At the product design level, the VDM code was refined into more concrete specifications. Finally, part of the lower-level code was formally specified and developed using CCS. As an outcome of the project, Praxis reported that no extra costs were incurred due to the use of formal methods. The productivity was comparable or better with respect to other projects that used only informal specification methods.

Some faults in the software were discovered that would have escaped traditional testing techniques. Overall, the quality of the produced software, in terms of number of faults per thousand lines of code, was estimated to be much higher, about two to ten times better than in comparable software for air traffic management systems that were not developed using formal methods techniques.

5.9.3 The Paris Métro Signaling System (SACEM software)

The SACEM system (Guiho and Hennebert, 1990) was jointly developed by GEC Alsthom, MATRA Transport, and CSEE (Compagnie de Signaux et Entreprises Électrique) in 1989. The system, responsible for controlling the RER commuter train system in Paris, was composed of embedded software and hardware, the software part being composed of about 21,000 lines of Modula-2 code, 63% of which was deemed to be safety critical. Part of the SACEM software, including both on-board and trackside components, was formally specified using the B language, and then formally verified, with assistance of partially automated proof generation and checking tools. The SACEM project is an example of successful "reverse engineering," in the sense that formal specification and verification were conducted after development of the code. The system was finally certified by the French railway authority. The project required a substantial investment in terms of manpower; it has been estimated that the overall effort was on the order of 100 man-years, about 1.5 times the effort required to develop the system itself. The effort was justified by the highly safety-critical nature of the system.

According to the project team, one of the most serious difficulties that had to be overcome was the communication problem between the signaling engineers and the experts in formal verification. To bridge this gap, a full specification in natural language was derived from the formal specification. A significant advantage was felt to be the additional clarifications and deep insight into the system that resulted from the use of formal techniques.

5.9.4 The Mondex Electronic Purse

Although an instance of a security-critical (rather than safety-critical) application, we mention the following as a significant example of the use of formal refinement in an industrial-scale application. The Mondex Electronic Purse (Stepney et al., 2000) is an electronic system for e-commerce, based on a smartcard, produced by the (former) NatWest Development Team. Each purse must ensure security of all its transactions (authentication, correct transfer, and no value loss in transactions) without resorting to a central controller. Formal methods were used to formally prove a refinement relation between a highly abstract, much more comprehensible model of the system, and a more concrete model, implementing a protocol for money transfer over an insecure and lossy medium. The abstract model was targeted exclusively at specifying the security properties of the system. The refinement relation between the abstract and the concrete models was formally specified and verified (mostly by hand) using the Z language and some related tools, extended with additional ad-hoc rules for refinement. Recently the relationship between the two models has been also verified using retrenchment (Banach et al., 2005), an extension of refinement targeted at providing a formal account of aspects that are not adequately dealt with by conventional refinement, in particular—the impossibility of refining infinite types into finite types (such as the finiteness of the sequence number in the Mondex protocol).

5.9.5 The Flight Warning Computer (FWC) of A330 and A340 Aircraft

An example where the application of formal methods did not prove very beneficial involved the formal specification of the Flight Warning Controller of the Airbus A330 and A340 aircraft (Garavel and Hautbois, 1993). The project was undertaken by Aerospatiale, INRIA, CEAT (the French certification authority), and British Aerospace. The FWC was designed to be a distributed system (composed of three CPUs and a software controller programmed in ADA, PL/M, and assembler), and responsible for monitoring different sources of input, checking for their validity, and raising suitable alarms whenever some conditions were not met. Once the implementation of the system was completed, it was decided to use it as a testbed for the application of formal methods, with the aim of estimating whether formal methods techniques could lead to a reduction in development costs. The formal specification was written in LOTOS, and it comprised over 27,500 lines of code. Unfortunately, some aspects of the system could not be adequately modeled in LOTOS, for instance, the real-time features of the system. Moreover, the formal specification had the same level of granularity of the corresponding ADA program, but was about 30% larger. The final specification was too large to analyze by existing formal verification tools and, as consequence, it was only validated by testing and simulation. Eventually, the LOTOS specification was translated into C code, the analysis of which revealed a number of errors in the LOTOS code. Finally, the formal specification was used to automatically generate executable code, but the resulting program was orders of magnitude slower than required by the real application. The FWC system is often cited as an example of unsuccessful application of formal methods, due to the choice of unrealistic goals and the use of unsuitable languages and techniques.

5.9.6 The Traffic Collision Avoidance System (TCAS)

Another example of the application of formal methods in the air traffic transport domain is the formal specification of the Traffic Collision and Avoidance System (TCAS) II, which is installed on all commercial aircraft with more than 30 seats (or less, depending on the regulation authority), and whose purpose is to reduce the chance of a midair collision. After some flaws were discovered in the original TCAS specification, a formal specification was commissioned to the Safety-Critical Systems Research Group led by Nancy Leveson (then at the University of California, and now at MIT). A specification in natural language was developed as well, in parallel with the formal specification, but is was soon abandoned due to its overwhelming complexity. The formal specification was written in the Requirements State Machine Language, based on a variant of Statecharts. The original specification did not undergo tool-supported formal analysis, but nevertheless it was deemed extremely useful by the project reviewers, in that it clarified the TCAS requirements and increased confidence in the original specification. Eventually, the formal specification was automatically checked for completeness and consistency (Heimdahl and Leveson, 1996), and provably-correct code generated from the RSML specification. The original specification is still being used to specify and evaluate potential upgrades to TCAS.

5.9.7 The Rockwell AAMP5 Microprocessor

In the area of hardware design, a project involving the use of formal methods was carried out to specify the AAMP5 microprocessor produced by Rockwell and formally verify its microcode (Miller and Srivas, 1995). The project was undertaken by Collins Commercial Avionics and SRI. The AAMP5 microprocessor has a complex architecture, designed for high-level languages such as ADA, and implements floating point arithmetic in the microcode. About half of the AAMP5 instructions were specified using the PVS theorem prover, and the microcode for some representative instructions was formally verified.

5.9.8 The VIPER Microprocessor

A more controversial example of the application of formal methods in the area of hardware design concerned the formal verification of the VIPER microprocessor (Cullyer, 1988). VIPER was a microprocessor with a simple architecture specifically developed for safety-critical applications. Formal methods were used throughout the development cycle of VIPER, at different levels and using different techniques. The work was performed by the Royal Signals and Radar Establishment (RSRE) with funding from the British Ministry of Defence. Parts of the system were specified and verified using the HOL theorem prover and the LCF-LSM language (a precursor of HOL). In particular, the proof relating the top-level specification of the microprocessor and an abstract view of its register transfer level was carried out in HOL. Some controversy arose in connection with some marketing claims, made by a commercial company licensing some aspects of VIPER, about the fact that the processor had gone through a complete formal verification, from the top specification down to the gate-level design, which was not the case. These claims were condemned in a review report on VIPER commissioned by NASA. Other controversy was due to the prohibitive costs of the formal development, and the fact that formal verification did not produce significant results, except from revealing some minor flaws in the specification, which were of no concern for the fabricators of the chip.

5.9.9 The SRT Division Algorithm

The correctness of the SRT floating point division algorithm was formally verified in 1995 (Clarke et al., 1996). The SRT algorithm is similar to the one used in the Pentium processor, whose bug caused a loss of nearly 500 million dollars in 1994. The SRT uses a radix-4 division algorithm, which was formalized as a set of algebraic relations over the real numbers. A correctness statement was defined and verified using theorem proving techniques. Later, in 1996, a generalization of the SRT division algorithm (for the general theory with generic radix) was again verified using the PVS theorem prover (Rue et al., 1996).

5.9.10 The IEEE Futurebus+ Cache Coherence Protocol

The IEEE Futurebus+ Cache Coherence Protocol, described in Standard 896.1-1991, was formally verified in 1992 (Clarke et al., 1993). The protocol ensures data consistency in hierarchical systems composed of multiple processors and caches. The protocol was formally specified in the SMV language, and then the SMV model checker was used to prove that the resulting transition system satisfied a property of cache coherence, expressed as a requirement in CTL temporal logic. The overall specification comprised about 2,300 lines of SMV code. A number of different flaws and potential errors, previously unnoticed, were discovered in the original protocol. Furthermore, the outcome of the project was a clear, unambiguous, and concise model of the protocol. This was the first attempt to validate the protocol using formal methods techniques, and it is often cited as one of the first significant achievements of model checking techniques in formal verification.

5.9.11 The BOS Control System

The BOS software (Beslin & Ondersteunend Systeem in Dutch) (Tretmans et al., 2001) is an automatic system aimed at protecting the harbor of Rotterdam from floodings, while at the same time guaranteeing ship traffic at all times, except under unacceptable weather conditions. BOS controls a movable barrier, taking decisions of when and how the barrier has to move, on the basis of a number of measures including water levels, tidal info, and weather predictions. BOS is a highly critical system (it is categorized in the highest Safety Integrity Level, according to IEC 61508). The design and implementation of BOS was undertaken by CMG Den Haag B.V. (Computer Management Group), in collaboration with the Formal Methods Group at the University of Twente. In the project, different methodologies were chosen to support system development. In particular, formal methods were used to assist the formal specification of the design and validating crucial parts of it. The design was formally specified in Promela (the language of the SPIN model checker) for the control part, and in Z for the data part. The formal validation of the design focused on the communication protocols between BOS and the environment. The final implementation was carried out using informal rules to map Z code to C++ statements. Although formal methods were used only in selected parts of the project, they proved very effective. In the initial stage of the project, they helped uncover several issues in an existing design of the protocols, convincing management to redesign the interface. Alternative protocols were then developed and formally validated. As a result of the application of formal methods, only a few faults, not impacting system reliability, were found during customer acceptance testing and in operation. Overall, the main benefit of formal methods was felt to be the improved quality of the final product, which justified the slightly increased costs of the project.

5.9.12 The SLAM Project and the Static Design Verifier

The SLAM project (SLAM, 2009) is an example of successful application of formal methods for checking the quality of software interfaces. In particular, the project, undertaken by Microsoft Research, focuses on the analysis of device drivers for the Windows^{TM} operating system, and their interaction with the API provided by the Windows kernel. In general, errors in the device drivers may have serious consequences, in that they may cause the operating system to crash or hang. The project is still ongoing. The history of the project, and pointers to several articles on the subject, can be found in Ball et al. (2004). The analysis engine developed within the SLAM project forms the basis of the Static Design Verifier (SDV) tool. SDV can be used to check compliance of a device driver against a set of rules formalizing proper interaction with the operating system kernel. The project leverages the use of several formal methods techniques, ranging from type checking to symbolic execution, model checking, static analysis, and theorem proving. In particular, SDV is based on three (iterative) steps and a verification strategy known as *counterexample-guided abstraction refinement* (see also Section 5.10). The first step consists of building an abstraction of the original program. The second step checks whether the abstracted program shows any error. The final step refines the abstraction in case of a false error path, and iterates the process. The SDV tool has been used to check the quality of several device drivers, both drivers developed by Microsoft and third-party drivers. Over the years, several hard-to-reproduce errors were uncovered in a number of drivers. The analyzed drivers had been in use for several years, and the errors had not been found using standard testing methodologies and code reviews. Most errors were acknowledged as such by the driver developers, whereas only a few of them were false errors, mostly due to deficiencies in the models or the rules used by SDV.

5.9.13 The Intel® Core™ i7 Processor Execution Cluster

Formal methods have been used in the verification of the Intel Core i7 processor (Kaivola et al., 2009). The Intel Core i7 is a multi-core processor following the IA-32 architectural model. In particular, formal methods were used for pre-silicon validation of the execution cluster EXE, a component that is responsible for carrying out data computations for all microinstructions. The EXE cluster is, in turn, composed of a few units, including integer and floating point units, units for memory address calculations, and interfaces. Overall, the cluster implements more than 2,700 microinstructions and supports multi-threading. The scope of formal methods was extended, with respect to previous projects (compare (Kaivola et al., 2009; Schubert, 2003), to include validation of the full datapath, the control logic, and the state components. Technically, formal methods based on symbolic simulation and manually generated inductive invariants were used in the validation process. A significant outcome of the project was that, for the first time, formal verification entirely replaced traditional coverage-driven testing. Results showed that formal verification was a viable and competitive alternative with respect to traditional testing-based methodologies, both in terms of time and costs, and in terms of the quality of the resulting design. As a matter of fact, the execution cluster had the lowest number of undetected errors escaping to silicon for any other cluster of the design. Experience gained in the project, at the time this book was written, is being exploited in further projects and is expected to show very effective validation results.

5.10 Conclusions and Future Directions

The use of formal methods to support system design and development is becoming more and more pervasive, and this trend is expected to increase in the next few decades, in particular in safety-critical domains. The reasons for this success are manifold. First, the growing complexity of safety-critical systems is making traditional development methodologies less reliable and more costly. Second, formal verification techniques have the potential to carry out extensive, if not exhaustive, verification analyses, which can be exploited to increase the level of confidence in the correctness, safety, and reliability properties of a given design. Finally, the use of formal methods is flexible, for instance, in terms of the level of formalization and extent of application; hence, it can be fruitfully adapted to the specific project at hand and used to integrate, not necessarily replace, more traditional design and verification methodologies.

Another reason for the growing success of formal methods is given by the progress in research and the experience in their use gained in recent years that have made them more practicable. Although the difficulties and limitations related to the application of formal methods should not be underestimated, it is true that recent years have seen formidable advances as regards the level of maturity, tool support, and cost-effectiveness of formal methods technologies (compare Section 5.9). In fact, many traditional arguments against the use of formal methods (Hall, 1990; Bowen and Hinchey, 1995) are contradicted by several hands-on experiences in industry. This is becoming recognized also by certification authorities, as we will see in Chapter 6.

In addition to bringing current state-of-the-art techniques into standard industrial practice, there are several research directions that are currently being explored and will likely attract more and more interest in forthcoming years. On the one hand, technological advances are being explored to improve the performance of existing techniques and make them suitable for industrial-size projects. On the other hand, improvements to current specification languages and verification techniques, as regards expressiveness, ease of use, and extent of application, are being pursued. We mention a few of these research directions, in particular in the field of model checking and formal safety analysis, that were the main focus of this chapter.

Several techniques are targeted at improving the performance and effectiveness of existing technologies, by scaling down the models that undergo formal verification. Examples of optimization techniques, already mentioned in Section 5.7.3, include symmetry reduction and partial-order reduction (Clarke et al., 2000; Baier and Katoen, 2008). Further techniques exploit different forms of model abstraction to reduce the dimension and complexity of models. As an example, **Counterexample Guided Abstraction Refinement (CEGAR)** (Clarke et al., 2003) is based on the idea of automatically constructing an abstraction of a model based on a set of predicates. The abstraction is safe; that is, if the abstract model can be proved correct with respect to a property, then the original model is also correct. If the abstraction is not correct, a counterexample is produced and checked against the original model. If it is not a counterexample for the original model, it is used to refine the set of predicates and the process is iterated.

Many practical systems can be conveniently described as infinite-state systems. A notable example involves modeling analog devices, such as mechanical, hydraulic, and pneumatic components, in terms of the physical laws that describe their behavior (e.g., by means of differential equations). A nontrivial challenge concerns the modeling and verification of systems that include both digital and analog components. A standard approach to address this issue is to discretize the continuous behavior of analog components. In fact, this was the approach we followed to encode the Three Mile Island example presented in this book. Alternatively, these systems could be formalized as **hybrid systems** (Alur et al., 1995; Henzinger, 1996; Lynch et al., 2003). A hybrid system can be described as an (infinite-state) machine that admits two different types of transitions: (1) discrete transitions, which are instantaneous and model discrete state changes, and (2) timed (continuous) transitions, which have a time duration and model continuous updates of physical quantities. The field of hybrid systems verification has received significant attention in recent years.

Concerning formal safety analysis, several research directions are being investigated. The generation of fault trees, for instance, is an open research area in different respects. To improve performance, novel algorithms for fault tree generation (Bozzano et al., 2007), based on Binary Decision Diagrams or Bounded Model Checking, or combinations of both technologies, are being investigated. A further research direction is the generation of dynamic fault trees (Dugan et al., 1992; Manian et al., 1998). Dynamic fault trees, for instance, can incorporate information concerning the duration and temporal ordering between events in the tree (Bozzano and Villafiorita, 2003; Åkerlund et al., 2006). Finally, another line of research is aimed at improving the logical structure of automatically generated trees, for reasons of readability or engineering semantics. A notable example includes techniques based on retrenchment (Banach and Bozzano, 2006) to generate hierarchical trees that are shaped according to the system structure. Finally, probabilistic model checking techniques (PRISM, 2009; MRMC, 2009) can be used to address questions that arise in probabilistic safety assessment, such as probabilistic evaluation of fault trees and system performability measures.

An area we have not discussed in this book concerns the application of formal methods to address the problem of human error and human-machine interface. We refer the reader to Leveson et al. (1997), Javaux and Olivier (2000), Degani and Heymann (2002), (Rushby 2002, 2001a,b), Lüdtke and Pfeiffer (2007) as a starting point on the subject.

References

Abdulla, P.A., J. Deneux, G. Stålmarck, H. Ågren, and O. Åkerlund (2004). Designing safe, reliable systems using scade. In T. Margaria, B. Steffen, A. Philippou, and M. Reitenspieß (Eds.), *Proc. 1st International Symposium on Leveraging Applications of Formal Methods (ISoLA 2004)*, Volume 4313 of *LNCS*, pp. 111–118. Springer.

Abrial, J.-R. (1996). *The B-Book: Assigning Programs to Meanings*. Cambridge U.K.: Cambridge University Press.

Abrial, J.-R., M. Lee, D.S. Neilson, P. Scharbach, and I.H. Sorensen (1991). The B-method. In S. Prehn and W. Toetenel (Eds.), *Proc. 4th International Symposium of VDM Europe (VDM'91)*, Volume 552 of *LNCS*, pp. 398–405. Springer.

Accellera (Last retrieved on November 15, 2009). Accellera. http://www.accellera. org.

ACL2 (Last retrieved on November 15, 2009). ACL2: A Computational Logic for Applicative Common Lisp. http://www.cs.utexas.edu/users/moore/acl2.

Åkerlund, O., P. Bieber, and E. Böede et al. (2006). ISAAC, a framework for integrated safety analysis of functional, geometrical and human aspects. In *Proc. European Congress on Embedded Real Time Software (ERTS 2006)*.

Alur, R., C. Courcoubetis, N. Halbwachs, et al. (1995). The algorithmic analysis of hybrid systems. *Theoretical Compututer Science* 1, 3–34.

Alur, R., T.A. Henzinger, and P.-H. Ho (1996). Automatic symbolic verification of embedded systems. *IEEE Transactions on Software Engineering* 22(3), 181–201.

Baeten, J.C.M. (2005). A brief history of process algebra. *Theoretical Computer Science* 335(2-3), 131–146.

Baier, C. and J.-P. Katoen (2008). *Principles of Model Checking*. Cambridge, MA: MIT Press.

Ball, T., B. Cook, V. Levin, and S.K. Rajamani (2004). SLAM and Static Driver Verifier: technology transfer of formal methods inside microsoft. In E.A. Boiten, J. Derrick, and G. Smith (Eds.), *Proc. 4th International Conference on Integrated Formal Methods (IFM 2004)*, Volume 2999 of *LNCS*, pp. 1–20. Berlin: Springer.

Banach, R. and M. Bozzano (2006). Retrenchment, and the generation of fault trees for static, dynamic and cyclic systems. In J.Górski (Ed.), *Proc. 25th International Conference on Computer Safety, Reliability, and Security (SAFECOMP 2006)*, Volume 4166 of *LNCS*, pp. 127–141. Springer.

Banach, R., M. Poppleton, C. Jeske, and S. Stepney (2005). Retrenching the purse: finite sequence numbers, and the tower pattern. In J. Fitzgerald, I. Hayes, and A. Tarlecki (Eds.), *Proc. International Symposium of Formal Methods Europe (FM 2005)*, Volume 3582 of *LNCS*, pp. 382–398. Springer.

Barroca, L.M. and J.A. McDermid (1992). Formal methods: Use and relevance for the development of safety-critical systems. *Computer Journal* 35(6), 579–599.

Beizer, B. (1990). *Software Testing Techniques*. New York: Van Nostrand Reinhold Co.

Beizer, B. (1995). *Black-Box Testing: Techniques for Functional Testing of Software and Systems*. New York: Wiley.

Bergstra, J. and J. Klop (1984). Process algebra for synchronous communication. *Information and Control* 60(1-3), 109–137.

Bieber, P., C. Castel, and C. Seguin (2002). Combination of fault tree analysis and model checking for safety assessment of complex system. In F. Grandoni and P. Thévenod-Fosse (Eds.), *Proc. 4th European Dependable Computing Conference (EDCC-4)*, Volume 2485 of *LNCS*, pp. 19–31. Berlin: Springer.

Biere, A., A. Cimatti, E.M. Clarke, and Y. Zhu (1999). Symbolic model checking without bdds. In R. Cleaveland (Ed.), *Proc. 5th International Conference on Tools and Algorithms for Construction and Analysis of Systems (TACAS'99)*, Volume 1579 of *LNCS*, pp. 193–207. Berlin: Springer.

Bowen, J.P. and M.G. Hinchey (1995). Seven more myths of formal methods. *IEEE Software* 12(4), 34–41.

Bowen, J.P. and V. Stavridou (1992). Safety-critical systems, formal methods and standards. *BCS/IEE Software Engineering Journal* 8(4), 189–209.

Bozzano, M., A. Cimatti, and F. Tapparo (2007). Symbolic fault tree analysis for reactive systems. In *Proc. 5th International Symposium on Automated Technology for Verification and Analysis (ATVA 2007)*, Volume 4762 of *LNCS*, pp. 162–176. Berlin: Springer.

Bozzano, M. and A. Villafiorita (2003). Integrating fault tree analysis with event ordering information. In *Proc. European Safety and Reliability Conference (ESREL 2003)*, pp. 247–254. Leiden, The Netherlands: Balkema Publisher.

Bozzano, M. and A. Villafiorita (2007). The FSAP/NuSMV-SA safety analysis platform. *Software Tools for Technology Transfer* 9(1), 5–24.

Bozzano, M., A. Villafiorita, and O. Åkerlund et al. (2003). ESACS: An integrated methodology for design and safety analysis of complex systems. In *Proc. European Safety and Reliability Conference (ESREL 2003)*, pp. 237–245. Leiden, The Netherlands: Balkema Publisher.

Bryant, R.E. (1992). Symbolic boolean manipulation with ordered binary decision diagrams. *ACM Computing Surveys* 24(3), 293–318.

Burch, J.R., E.M. Clarke, D. Long, K.L. McMillan, and D.L. Dill (1994). Symbolic model checking for sequential circuit verification. *IEEE Transactions on Computer-Aided Design of Integrated Circuits and Systems* 13(4), 401–424.

Burch, J.R., E.M. Clarke, K.L. McMillan, D.L. Dill, and L.J. Hwang (1992). Symbolic model checking: 10^{20} states and beyond. *Information and Computation* 98(2), 142–170.

Cardelli, L. and A. Gordon (2000). Mobile ambients. *Theoretical Computer Science* 240, 177–213.

Chaochen, Z., C.A.R. Hoare, and A.P. Ravn (1991). A calculus of duration. *Information Processing Letters* 40(5), 269–276.

Cimatti, A., E.M. Clarke, F. Giunchiglia, and M. Roveri (2000). NuSMV: A new symbolic model checker. *Software Tools for Technology Transfer* 2(4), 410–425.

Cimatti, A., E.M. Clarke, and E. Giunchiglia et al. (2002). NuSMV2: An opensource tool for symbolic model checking. In E. Brinksma and K. Larsen (Eds.), *Proc. 14th International Conference on Computer Aided Verification (CAV'02)*, Volume 2404 of *LNCS*, pp. 359–364. Berlin: Springer.

Clarke, E.M. and E.A. Emerson (1981). Synthesis of synchronization skeletons for branching time temporal logic. In D. Kozen (Ed.), *Proc. Workshop on Logic of Programs*, Volume 131 of *LNCS*, pp. 52–71. Berlin: Springer.

Clarke, E.M., E.A. Emerson, and A. Sistla (1986). Automatic verification of finite-state concurrent systems using temporal logic specifications. *ACM TOPLAS* 8(2), 244–263.

Clarke, E.M., S.M. German, and X. Zhao (1996). Verifying the SRT division algorithm using theorem proving techniques. In R. Alur and T.A. Henzinger (Eds.), *Proc. 8th International Conference on Computer Aided Verification (CAV'96)*, Volume 1102 of *LNCS*, pp. 111–122. Berlin: Springer.

Clarke, E.M., O. Grumberg, H. Hiraishi, S.K. Jha, D. Long, K.L. McMillan, and L. Ness (1993). Verification of the FUTUREBUS+ cache coherence protocol. In *Proc. International Symposium on Computer Hardware Description Languages and their Applications (CHDL-93)*. Amsterdam: Elsevier.

Clarke, E.M., O. Grumberg, S. Jha, Y. Lua, and H. Veith (2003). Counterexample-guided abstraction refinement for symbolic model checking. *Journal of the ACM* 50(5), 752–794.

Clarke, E.M., O. Grumberg, and D.A. Peled (2000). *Model Checking*. Cambridge, MA: MIT Press.

Clarke, E.M. and J.M. Wing (1996). Formal methods: State of the art and future directions. *ACM Computing Surveys* 28, 626–643.

Coq (Last retrieved on November 15, 2009). The Coq proof assistant. `http://coq.inria.fr`.

Coudert, O. and J.C. Madre (1993). Fault tree analysis: 10^{20} prime implicants and beyond. In *Proc. Annual Reliability and Maintainability Symposium (RAMS'93)*, pp. 240–245. Washington, D.C.: IEEE Computer Society.

Cullyer, W. (1988). Implementing safety critical systems: The VIPER microprocessor. In *Proc. VLSI Specification, Verification and Synthesis*, pp. 1–26. Dordrecht, The Netherlands: Kluwer Academic.

Craigen, D. and K. Summerskill (Ed.) (1989). *Proc. FM 89: A Workshop on the Assessment of Formal Methods for Trustworthy Computer Systems*. Washington, D.C.: IEEE Computer Society.

Daws, C. and S. Yovine (1995). Two examples of verification of multirate timed automata with KRONOS. In *Proc. 16th IEEE Real-Time Systems Symposium*, pp. 66–75. Washington, D.C.: IEEE Computer Society.

Degani, A. and M. Heymann (2002). Formal verification of human-automation interaction. *Human Factors* 44(1), 28–43.

Deneux, J. and O. Åkerlund (2004). A common framework for design and safety analyses using formal methods. In *Proc. International Conference on Probabilistic Safety Assessment and Management (PSAM7/ESREL'04)*.

Dijkstra, E.W. (1976). *A Discipline of Programming*. Upper Saddle River, NJ: Prentice Hall.

Dill, D.L., A.J. Drexler, A.J. Hu, and C.H. Yang (1992). Protocol verification as a hardware design aid. In *Proc. IEEE 1992 International Conference on Computer Design, VLSI in Computers and Processors*, pp. 522–525. Washington, D.C.: IEEE Computer Society.

Dugan, J., S. Bavuso, and M. Boyd (1992). Dynamic fault-tree models for fault-tolerant computer systems. *IEEE Transactions on Reliability* 41(3), 363–77.

E (Last retrieved on November 15, 2009). The E Equational Theorem Prover. `http://www4.informatik.tu-muenchen.de/~schulz/WORK/eprover.html`.

Eèn, N. and N. Sörensson (2003). An extensible SAT solver. In E. Giunchiglia and A. Tacchella (Eds.), *Proc. Conference on Theory and Applications of Satisfiability Testing (SAT 2003)*, Volume 2919 of *LNCS*, pp. 502–518. Berlin: Springer.

Emerson, E.A. (1990). Temporal and modal logic. In J. van Leeuwen (Ed.), *Handbook of Theoretical Computer Science*, Volume B, pp. 995–1072. Amsterdam: Elsevier Science.

ESACS (Last retrieved on November 15, 2009). The ESACS Project. `http://www.esacs.org`.

FSAP (Last retrieved on November 15, 2009). The FSAP/NuSMV-SA platform. `https://es.fbk.eu/tools/FSAP`.

Futatsugi, K., J. Goguen, J.-P. Jouannaud, and J. Meseguer (1985). Principles of OBJ2. In *Proc. 12th ACM SIGACT-SIGPLAN Symposium on Principles of Programming Languages (POPL'85)*, pp. 52–66. New York: ACM.

Gallier, J.H. (1986). *Logic for Computer Science: Foundations of Automatic Theorem Proving*. New York: Wiley.

Garavel, H. and R. Hautbois (1993). An experience with the LOTOS formal description technique on the flight warning computer of the Airbus A330/340 aircrafts. In *1st AMAST International Workshop on Real-Time Systems*. Berlin: Springer.

Goguen, J., C. Kirchner, H. Kirchner, A. Megrelis, J. Meseguer, and T. Winkler (1988). An introduction to OBJ 3. In *Proc. 1st International Workshop on Conditional Term Rewriting Systems*, Volume 308 of *LNCS*, pp. 258–263. Berlin: Springer.

Grumberg, O. and H. Veith (Eds.) (2008). *25 Years of Model Checking: History, Achievements, Perspectives*, Volume 5000 of *LNCS*. Berlin: Springer.

Guiho, G.D. and C. Hennebert (1990). SACEM software validation. In *12th International Conference on Software Engineering*, pp. 186–191. Washington, D.C.: IEEE Computer Society.

Gupta, A., E.M. Clarke, and O. Strichman (2004). SAT based counterexample-guided abstraction refinement. *IEEE Transactions on Computer Aided Design* 23, 1113–1123.

Guttag, J.V. and J.J. Horning (1993). *Larch: Languages and Tools for Formal Specification*. Springer. Written with S.J. Garland, K.D. Jones, A. Modet, and J.M. Wing. New York: Springer-Verlag.

Hall, A. (1990). Seven myths of formal methods. *IEEE Software* 7(5), 11–19.

Hall, A. (1996). Using formal methods to develop an ATC information system. *IEEE Software* 13(2), 66–76.

Harel, D. (1987). Statecharts: A visual formalism for complex systems. *Science of Computer Programming* 8, 231–274.

Heimdahl, M. and N.G. Leveson (1996). Completeness and consistency in hierarchical state-based requirements. *IEEE Transactions on Software Engineering* 22(6), 363–377.

Heljanko, K., T. Junttila, and T. Latvala (2005). Incremental and complete bounded model checking for full PLTL. In K. Etessami and S.K. Rajamani (Eds.), *Proc. 17th International Conference on Computer Aided Verification (CAV'05)*, Volume 3576 of *LNCS*, pp. 98–111. Berlin: Springer.

Henzinger, T.A. (1996). The theory of hybrid automata. In *Proc. Symposium on Logic in Computer Science (LICS'96)*, pp. 278–292. Washington, D.C.: IEEE Computer Society.

Hoare, C.A.R. (1969). An axiomatic basis of computer programming. *Communications of the ACM* 12(10), 576–580.

Hoare, C.A.R. (1985). *Communicating Sequential Processes*. Upper Saddle River, NJ: Prentice Hall.

HOL (Last retrieved on November 15, 2009). The HOL Interactive Proof Assistant for Higher Order Logic. http://hol.sourceforge.net.

Houston, I. and S. King (1991). CICS project report: Experiences and results from the use of Z in IBM. In S. Prehn and W. Toetenel (Eds.), *Proc. 4th International Symposium of VDM Europe (VDM'91)*, Volume 552 of *LNCS*, pp. 588–596. Berlin: Springer.

IEEE 1850 (Last retrieved on November 15, 2009). IEEE 1850. http://www.eda.org/ieee-1850.

ISAAC (Last retrieved on November 15, 2009). The ISAAC Project. http://www.cert.fr/isaac.

Jackson, D. (2006). *Software Abstractions: Logic, Language, and Analysis*. Cambridge, MA: MIT Press.

Javaux, D. and E. Olivier (2000). Assessing and Understanding Pilots' Knowledge of Mode Transitions on the A340-200/300. In K. Abbott, J.-J. Speyer, and G. Boy (Eds.), *Proc. International Conference on Human-Computer Interaction in Aeronautics (HCI-Aero 2000)*, pp. 163–168. Toulouse, France: Cépaduès-Ed.

Jones, C.B. (1990). *Systematic Software Development Using VDM*. Upper Saddle River, NJ: Prentice Hall.

Joshi, A. and M. Heimdahl (2005). Model-based safety analysis of simulink models using SCADE design verifier. In R. Winther, B. Gran, and G. Dahll (Eds.), *Proc. 24th International Conference on Computer Safety, Reliability and Security (SAFECOMP 2005)*, Volume 3688 of *LNCS*, pp. 122–135. Berlin: Springer.

Joshi, A., S. Miller, M. Whalen, and M. Heimdahl (2005). A proposal for model-based safety analysis. In *Proc. 24th Digital Avionics Systems Conference (DASC 2005)*. Washington, D.C.: IEEE Computer Society.

Kaivola, R., R. Ghughal, and N. Narasimhan et al. (2009). Replacing testing with formal verification in Intel® Core™ i7 processor execution engine validation. In A. Bouajjani and O. Maler (Eds.), *Proc. 21st International Conference on Computer Aided Verification (CAV'09)*, Volume 5643 of *LNCS*, pp. 414–429. Berlin: Springer.

Kaner, C., H.Q. Nguyen, and J.L. Falk (1993). *Testing Computer Software*. New York: Wiley.

Leveson, N.G., L. Pinnell, S. Sandys, S. Koga, and J. Reese (1997). Analysing software specifications for mode confusion potential. In C.W. Johnson (Ed.), *Proc. Workshop on Human Error and System Development*, pp. 132–146.

LOTOS (1989). Information Processing Systems, Open Systems Interconnection, LOTOS—A Formal Description Technique Based on the Temporal Ordering of Observational Behaviour. Volume IS-8807 of *International Standards*. ISO.

Lüdtke, A. and L. Pfeiffer (2007). Human error analysis based on a semantically defined cognitive pilot model. In F. Saglietti and N. Oster (Eds.), *Proc. 26th International Conference on Computer Safety, Reliability and Security (SAFECOMP 2007)*, Number 4680 in LNCS, pp. 134–147. Berlin: Springer.

Lynch, N.A., R. Segala, and F.W. Vaandrager (2003). Hybrid I/O automata. *Information and Computation* 185(1), 105–157.

Manian, R., J. Dugan, D. Coppit, and K. Sullivan (1998). Combining various solution techniques for dynamic fault tree analysis of computer systems. In *Proc. 3rd IEEE International*

Symposium on High-Assurance Systems Engineering (HASE '98), pp. 21–28. Washington, D.C.: IEEE Computer Society.

McMillan, K.L. (1993). *Symbolic Model Checking*. Dordrecht, The Netherlands: Kluwer Academic.

Miller, S.P. and M. Srivas (1995). Formal verification of the AAMP5 microprocessor: A case study in the industrial use of formal methods. In *Proc. Workshop on Industrial-Strength Formal Specification Techniques (WIFT'95)*, pp. 2–16. Washington, D.C.: IEEE Computer Society.

Milner, R. (1980). A calculus of communicating systems. Volume 92 of *LNCS*. Berlin: Springer.

Milner, R., J. Parrow, and D. Walker (1992). A calculus of mobile processes. *Information and Computation* 100, 1–77.

Mossakowski, T., A. Haxthausen, D. Sannella, and A. Tarlecki (2003). CASL—The Common Algebraic Specification Language: semantics and proof theory. *Computing and Informatics* 22, 285–321.

Mossakowski, T., A. Haxthausen, D. Sannella, and A. Tarlecki (2008). CASL, the Common Algebraic Specification Language. In D. Bjørner and M. Henson (Eds.), *Logics of Formal Specification Languages*, pp. 241–298. Berlin: Springer.

Moszkowski, B.C. and Z. Manna (1983). Reasoning in interval temporal logic. In *Proc. Workshop on Logic of Programs*, Volume 164 of *LNCS*, pp. 371–382. Berlin: Springer.

MRMC (Last retrieved on November, 15, 2009). MRMC: Markov Reward Model Checker. http://www.mrmc-tool.org.

Myers, G.J. (2004). *The Art of Software Testing*. New York: Wiley.

NuPRL (Last retrieved on November 15, 2009). The PRL Automated Reasoning Project. http://www.cs.cornell.edu/Info/Projects/NuPRL.

NuSMV (Last retrieved on November 15, 2009). NuSMV: A New Symbolic Model Checker. http://nusmv.fbk.eu.

Otter (Last retrieved on November 15, 2009). The Otter Theorem Prover. http://www.cs.unm.edu/~mccune/otter.

Peikenkamp, T., E. Böede, I. Brückner, H. Spenke, M. Bretschneider, and H.-J. Holberg (2004). Model-based safety analysis of a flap control system. In *Proc. 14th Annual International Symposium INCOSE 2004*.

Peterson, J. (1981). *Petri Net Theory and the Modeling of Systems*. Upper Saddle River, NJ: Prentice Hall.

Petri, C. (1966). Communication with Automata. DTIC Research Report AD0630125. Defense Technical Information Center, Fort Belvoir, VA.

Pill, I., S. Semprini, R. Cavada, M. Roveri, R. Bloem, and A. Cimatti (2006). Formal analysis of hardware requirements. In E. Sentovich (Ed.), *Proc. 43rd Design Automation Conference (DAC'06)*, pp. 821–826. New York: ACM.

Pnueli, A. (1981). A temporal logic of concurrent programs. *Theoretical Computer Science* 13, 45–60.

Prasad, M.R., A. Biere, and A. Gupta (2005). A survey of recent advances in SAT-based formal verification. *Software Tools for Technology Transfer* 7(2), 156–173.

PRISM (Last retrieved on November 15, 2009). The PRISM Probabilistic Model Checker. http://www.prismmodelchecker.org.

Prover9 (Last retrieved on November 15, 2009). The Prover9 Theorem Prover. http://www.cs.unm.edu/~mccune/prover9.

PSL (Last retrieved on November 15, 2009). The PSL/Sugar Consortium. http://www.pslsugar.org.

PVS (Last retrieved on November 15, 2009). The PVS Specification and Verification System. http://pvs.csl.sri.com.

Queille, J.-P. and J. Sifakis (1982). Specification and verification of concurrent systems in CÆSAR. In *Proc. International Symposium on Programming: 5th Colloquium (ISP'82)*, Volume 137 of *LNCS*, pp. 337–351. Berlin: Springer.

RAT (Last retrieved on November 15, 2009). The RAT Requirements Analysis Tool. `http://rat.fbk.eu`.

Rauzy, A. (1993). New algorithms for fault trees analysis. *Reliability Engineering and System Safety* 40(3), 203–211.

Rue, H., N. Shankar, and M.K. Srivas (1996). Modular verification of SRT division. In R. Alur and T.A. Henzinger (Eds.), *Proc. 8th International Conference on Computer Aided Verification (CAV'96)*, Volume 1102 of *LNCS*, pp. 123–134. Berlin: Springer.

Rushby, J. (1993). Formal Methods and the Certification of Critical Systems. Technical Report CSL-93-7, Menlo Park, CA: SRI.

Rushby, J. (2001a). Analyzing cockpit interfaces using formal methods. *Electronic Notes in Theoretical Computer Science* 43, 1–14.

Rushby, J. (2001b). Modelling the human in human factors. In U. Voges (Ed.), *Proc. 20th International Conference on Computer Safety, Reliability, and Security (SAFECOMP 2001)*, Volume 2187 of *LNCS*, pp. 86–91. Berlin: Springer.

Rushby, J. (2002). Using model checking to help discover mode confusions and other automation surprises. *Reliability Engineering and System Safety* 75(2), 167–177.

Schubert, T. (2003). High level formal verification of next-generation microprocessors. In *Proc. 40th Design Automation Conference (DAC'03)*, pp. 1–6. New York: ACM.

Setheo (Last retrieved on November 15, 2009). The Theorem Prover Setheo. `http://www.tcs.informatik.uni-muenchen.de/~letz/TU/setheo`.

Sheeran, M., S. Singh, and G. Stalmarck (2000). Checking safety properties using induction and a SAT-solver. In W.A. Hunt Jr. and S.D. Johnson (Eds.), *Proc. 3rd International Conference on Formal Methods in Computer-Aided Design (FMCAD 2000)*, Volume 1954 of *LNCS*, pp. 108–125. Berlin: Springer.

SLAM (Last retrieved on November 15, 2009). The SLAM Project. `http://research.microsoft.com/en-us/projects/slam`.

SPASS (Last retrieved on November 15, 2009). SPASS: An Automated Theorem Prover for First-Order Logic with Equality. `http://www.spass-prover.org`.

SPIN (Last retrieved on November 15, 2009). On-the-fly, LTL Model Checking with SPIN. `http://spinroot.com`.

Spivey, M.J. (1992). *The Z Notation: A Reference Manual.* Upper Saddle River, NJ: Prentice Hall.

STeP (Last retrieved on November 15, 2009). The Stanford Temporal Prover. `http://www-step.stanford.edu`.

Stepney, S., D. Cooper, and J. Woodcock (2000). An Electronic Purse: Specification, Refinement and Proof. Technical Report PRG-126, Oxford University Computing Laboratory.

Storey, N. (1996). *Safety Critical Computer Systems.* Reading, MA: Addison-Wesley.

Tretmans, J., K. Wijbrans, and M. Chaudron (2001). Software engineering with formal methods: The development of a storm surge barrier control system revisiting seven myths of formal methods. *Formal Methods in System Design* 19(2), 195–215.

UPPAAL (Last retrieved on November 15, 2009). UPPAAL. `http://www.uppaal.com`.

Vampire (Last retrieved on November 15, 2009). Vampire. `http://www.voronkov.com/vampire.cgi`.

Vardi, M.Y. (2001). Branching vs. linear time: final showdown. In T. Margaria and W. Yi (Eds.), *Proc. 7th International Conference on Tools and Algorithms for the Construction and Analysis of Systems (TACAS'01)*, Volume 2031 of *LNCS*, pp. 1–22. Berlin: Springer.

Vardi, M.Y. and P. Wolper (1986). An automata-theoretic approach to automatic program verification. In *Proc. Symposium on Logic in Computer Science (LICS '86)*, pp. 332–344. Washington, D.C.: IEEE Computer Society.

Vesely, W., M. Stamatelatos, J. Dugan, J. Fragola, J. Minarick III, and J. Railsback (2002). Fault Tree Handbook with Aerospace Applications. Technical report, NASA.

Vesely, W.E., F.F. Goldberg, N.H. Roberts, and D.F. Haasl (1981). Fault Tree Handbook. Technical Report NUREG-0492, Systems and Reliability Research Office of Nuclear Regulatory Research, U.S. Nuclear Regulatory Commission.

VIS (Last retrieved on November 15, 2009). VIS: Verification Interacting with Synthesis. `http://vlsi.colorado.edu/~vis`.

Wiegers, K. (2001). Inspecting requirements. *StickyMinds Weekly Column*.

Wing, J. (1998). A symbiotic relationship between formal methods and security. In *Proc. Workshop on Computer Security, Dependability, and Assurance: From Needs to Solutions*, pp. 26–38. Washington, D.C.: IEEE Computer Society.

Chapter 6

Formal Methods for Certification

In Chapter 4 we described how the certification process integrates with the other development activities and why certification-related activities are better started from the early stages of system development. The process described there, however, is generic and high level. This chapter details requirements, steps, and procedures for system certification; describes various widely adopted standards and recommendations; and illustrates the position of standardization documents with respect to the usage of formal methods.

We begin by describing the certification process in avionics (Section 6.1), continue with an overview of various standards (Sections 6.2 to Section 6.8), and conclude with some considerations related to the use of formal methods for certification (Section 6.9).

6.1 Certification of Avionic Systems

In general, certification is the process by which a system is demonstrated to comply with a set of regulations and standards that defines the minimum requirements a system must have to be safely deployed, operated, and—at the end of its life cycle—dismissed. In avionics, certification is the process of determining airworthiness of aircraft, that is, the fact that the airplane can be safely flown. Airworthiness is determined by national and international certification authorities such as the FAA (Federal Aviation Authority) in the United States and EASA in Europe[1].

Different types of certificates exist, according to the purpose for which the aircraft is being certified. The FAA, for instance, distinguishes between *standard certificates*

---

[1] EASA replaced, at the end of 2009, the former European certification authority, JAA.

and *special airworthiness certificates.* Simplifying a bit, the first is used for standard operations, whereas the second grants special permissions, for example, those necessary to fly experimental aircraft or for testing aircraft under development.

The award of a standard certificate for an airplane requires the aircraft to have been granted a *Type Certificate* (TC). Type certification demonstrates the compliance of the design and construction with a set of process and technical requirements such as those mentioned in Federal Aviation Administration (2007), AIA, GAMA, and FAA Aircraft Certification Service (2004), Department of Defense (2005), and U.S. Government (2009).

The requirements are meant to demonstrate that the airplane can be safely flown and that it complies with other constraints, such as level of emission and noise. They are expressed in text and can be quite broad in scope, such as, for instance, 23.143, Part 23, of U.S. Government (2009):

(a) The airplane must be safely controllable and maneuverable during all flight phases including
1. Takeoff
2. Climb
3. Level flight
4. Descent
5. Go-around
6. Landing (power on and power off) with the wing flaps extended and retracted

(b) It must be possible to make a smooth transition from one flight condition to another (including turns and slips) without danger of exceeding the limit load factor, under any probable operating condition (including, for multi-engine airplanes, those conditions normally encountered in the sudden failure of any engine).

(c) If marginal conditions exist with regard to required pilot strength, the control forces necessary must be determined by quantitative tests.

or rather more specific, such as 6.2.2.22 of Department of Defense (2005):

Verify that control laws redundancy and failure management designs are safely implemented.

In demonstrating compliance of a design with the technical requirements, there are two main sources of technical difficulties that must be properly addressed. The first concerns the interpretation of requirements and the means of compliance suggested by the certification authority. The second concerns how to demonstrate compliance with the requirements when new technologies, procedures, and development methodologies are being deployed. Let us look in more detail at both of them.

The first issue, interpretation and means of compliance, is typically addressed by clarifications. The FAA, for instance, publishes Advisory Circulars, that is, documents

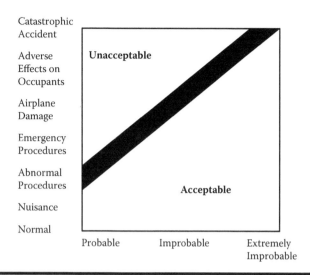

Figure 6.1 Probability of failure condition.

that clarify the interpretation of requirements and illustrate ways in which compliance can be demonstrated. As an example of an Advisory Circular, see Federal Aviation Administration (1998), which describes various means of complying with requirements 24.1309(b), (c), and (d), that requires one to demonstrate the "inverse relationship between the probability and the severity of each failure condition," as shown by the diagram of Figure 6.1.

The circular, although more than 10 years old, describes both quantitative and qualitative methods that are still in use. We list, in particular, the qualitative techniques that can be applied during design to demonstrate compliance:

- **Design Appraisal.** A qualitative appraisal of the integrity and safety of the design, based on experienced judgment and with emphasis on failure conditions that could prevent safe flight and landing.
- **Installation Appraisal.** A qualitative appraisal of the integrity and safety of the installation, based on experienced judgment and with emphasis on failure conditions that could prevent safe flight and landing.
- **Failure Mode and Effect Analysis,** which we described in Section 3.2.2.
- **Fault Tree or Reliability Block Diagram Analysis,** which we described in Section 3.2.1.

The document also presents a flowchart, "for the use of applicants who are not familiar with the various methods and procedures generally used by industry to conduct design safety assessment." It is shown in Figure 6.2.

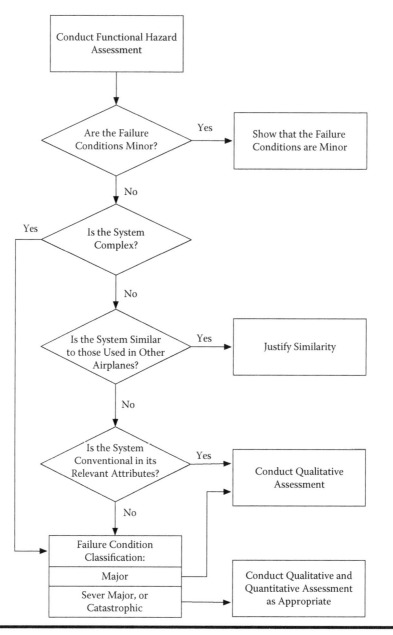

Figure 6.2 Methods and procedures for conducting safety design assessment.

The second issue deals with the so-called "special conditions," that is, particular means that are granted to demonstrate compliance with the standards. The special conditions are negotiated with the certification authority. Typically they concern the use of new technologies, either used during development or implemented by the system, that are not adequately covered by the standards. Granting a special condition often triggers an update of the standards, in order to adequately cover subsequent developments.

For instance, Spitzer (2006) reports that for the certification of the fly-by-wire system of the Airbus A3*0 family, various innovative topics had to be addressed with special conditions for certification:

- Flight envelope protection
- Side-stick controller
- Static stability
- Interaction of system and structure
- System safety assessment
- Lightning indirect effect and electromagnetic interference
- Electrical power
- Software verification and documentation, automatic code generation
- System validation
- Application-specific integrated circuit

In the remainder of the chapter we take a closer look at some of the standards adopted for the development of safety-critical applications. We then move on to describe the role that formal methods might have in implementing such standards. We primarily focus on software development and aerospace systems, although, in general, the process can be easily extended to other standards.

6.2 So Many Standards, So Little Time

Providing a comprehensive view of certification and available standards is a daunting task, as also noted by Sarah Sheard (2001), from which we quote:

> "Various combinations of national and international standards bodies, government organizations, professional societies, and other quasi legislative bodies—each vying for influence over a particular market sector— have promulgated a dizzying array of software and system process standards, recommended practices, guidelines, maturity models, and other frameworks."

One reason for the flourishing is that standardization activities often began through noncoordinated efforts conducted at the national level. A second reason is that standards (not necessarily safety standards) have been—in some cases, might still be—a way to protect national markets and build barriers to the entrance. A third reason is that new standards are defined to meet new issues and opportunities arising from changes in technologies and the experience gained in specific sectors.

Finally, to make the matter even more complex, we need to consider that standards have a life cycle, might derive from other standards, and might be endorsed by different regulatory entities, sometimes even using different identification systems. For instance, IEC-12207 ("Standard for Information Technology-Software Life Cycle Processes") is also a standard endorsed or published by the ISO, IEC, and EIA, under the same number. The standard derives from MIL-498B, which is now obsolete; MIL-498B , in turn, was introduced to replace three other military standards: DOD-STD-2167A, DOD-STD-7935A, and DOD-STD-1703. (See Appendix C for a list of various standardization bodies.)

Independent of the reasons that caused so many standards to be now available, some common aspects include

- **Prescription level.** Some documents, such are those defined by the FAA, are compulsory: compliance must be demonstrated in order to obtain certification. Other standards come in the form of recommendations: They describe best practices used to guarantee adequate levels of safety and quality; it is up to the production team to choose whether or not to apply them. Often, however, the distinction gets blurred, as some recommendations encode well-established practices, and any deviation from the recommendation needs to be carefully studied and motivated.
- **Reference sector.** Different markets might have different (safety) requirements, different traditions, and different engineering practices and constraints. Standards reflect such variety. Sticking to the safety-critical sector, we can list different standards for space, avionics, railway, maritime, nuclear energy production, and military, to name a few.
- **Scope.** Standards reflect and take into account different characteristics of systems. We can therefore find standards for software, hardware, and complex systems. Another distinction quite commonly found is what part of the development process a particular standard applies. Thus, together with standards defining the technical requirements for certification, we have standards and recommendations describing procedures and steps that must be followed for development, safety assessment, storage, quality, and management, to name a few.

Table 6.1 provides a classification of some well-known standards, according to reference sector and scope. The rows identify the target of the standard. We distinguish whether the standard mainly focuses on safety assessment, system development, or certification. The columns define the target reference sectors. We distinguish, in particular, avionics, railway, military. (Notice, however, that the classification is arbitrary because standards might apply to different sectors; military standards,

Table 6.1 Some Reference Standards in Different Application Sectors

	Avionics	Railway	Military	Other
Safety Assessment	ARP4761	IEC 50126	MIL-882B / DEF 00-56	IEC-61508 / CMMI+SAFE
Development Process	ARP4754 / DO-178B / ED-12B / DO-254	IEC 50129 / IEC 50128 / RSSB's Yellow Book	MIL-498 / DEF 00-55	IEEE12207 / ECSS / CMMI
System Certification	ARP4754		MIL-HNBK-516B	

for instance, include different civil areas, such as ground transportation, maritime, space, avionics, and various military areas, such as safe storage of ammunition.) The last column, finally, includes standards from other sectors, most notably electronics and space. Finally, at the intersection of rows and columns we list the standards.

In the remainder of this chapter we focus on a small subset of standards, that is, those that are grayed in Table 6.1 (Notice that CMMI and CMMI+SAFE were briefly discussed in Chapter 4). Concerning the others, here it suffices to say that a lot of information is available on the Internet. Wikipedia, for instance, provides a very good introduction to all of them. Moreover, standards published by public bodies, such as MIL (U.S. military), DEF (U.K. military), and ECSS (European Space Agency standards) are available for download, respectively, from ASSIST (Acquisition Streamlining and Standardization Information System) (Department of Defense, 2009); the U.K. Defence Standardization web site (Ministry of Defence, 2009); and from the European Cooperation for Space Standardization web site (European Cooperation for Space Standardization, 2009).

6.3 The ECSS System of Standards

Although they do not have a legal standing by themselves, the ECSS standards provide a very comprehensive system that covers all aspects of space engineering. Several ECSS standards, however, refer to general practices (e.g., project management) and are applicable to other domains. Here we just present their scope and organization.

The standards are organized in three main areas, identified by a letter:

- **Space project management branch,** identified by the letter **M**, and covering all aspects related to project management
- **Space Product Assurance,** identified by the letter **Q**, and covering all aspects related to quality assurance, including system safety
- **Space Engineering branch,** identified by the letter **E**, and covering all aspects related to system engineering, including software development

(A fourth area, **S**, is used to describe the actual organization of the standards and their usage.)

Each area has various documents, identified by a number, and describing specific disciplines. For instance, the number **10** is the Space Engineering area and it identifies the standards for "System Engineering." (ECSS disciplines can also be supported by handbooks and technical memoranda (ECSS Secreteriat, ESA-ESTEC, 2008).) A final letter is used to identify different releases of the same document; so, for instance, **E-ST-70C** is the third release of the engineering document **70** ("Ground systems and operations").

Figure 6.3, taken from ECSS Secreteriat, ESA-ESTEC (2008), shows the high-level organization of standards. Finally, notice that, at the time of writing, the standards are undergoing a major restructuring; the figure shows the situation after the batch 2 release.

6.4 Avionics Reference Standards

The reference guidelines for system development in the avionics sector are defined and published by a joint effort of EUROCAE (European Organization for Civil Aviation Equipment) (EUROCAE, 2009) and SAE (Society for Automotive Engineers) (SAE, 2009). Some widely adopted standards and recommendations in the avionics sector for system development and safety assessment include:

- **ARP4754/ED-79 Certification Considerations for Highly-Integrated or Complex Aircraft Systems** (SAE, 1996a) describes the system development process for complex and highly integrated systems.
- **ARP4761 Guidelines and Methods for Conducting the Safety Assessment Process on Civil Airborne Systems and Equipment** (SAE, 1996b) describes

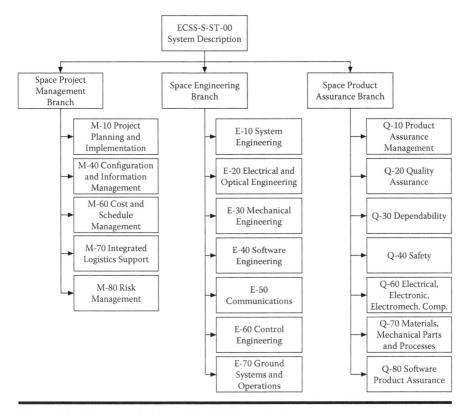

Figure 6.3 **The organization of the ECSS standards.**

the safety assessment process and techniques for complex highly integrated systems.

■ **DO-178B/ED-12B Software Considerations in Airborne Systems and Equipment Certification** (EUROCAE, 1992) describes best practices for the development of critical software components (see Advisory Circular (Federal Aviation Administration, 1993), for the note suggesting its usage).

■ **DO-254/ED-80 Design Assurance Guidance for Airborne Electronic Hardware** (RTCA, 2000) describes best practices for the development of hardware performing critical functions (see Advisory Circular (Federal Aviation Administration, 2005) for the note suggesting its usage).

The standards cover different aspects of system development, as shown in Figure 6.4. In particular, the ARPs cover aspects related to system development and integration, while the DOs cover aspect related to the implementation of single hardware and software items.

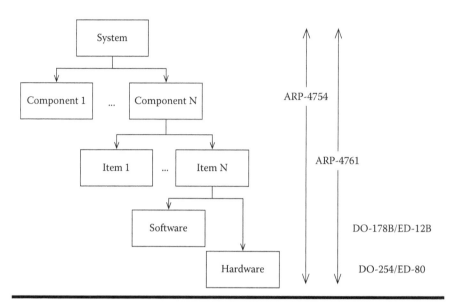

Figure 6.4 The role of some avionics standards in system engineering.

6.5 ARP 4754

ARP 4754 addresses the certification aspects of systems that implement aircraft-level functions, such as a fly-by-wire system. The document covers the aspects related to system development and integration. In the words of the authors:

> "These guidelines are intended to provide designers, manufacturers, installers, and certification authorities a common international basis for demonstrating compliance with airworthiness requirements applicable to highly-integrated or complex systems." (SAE, 1996a).

The document focuses on the main aspects characterizing the development and certification of a complex system, to include

- **System development,** defining a top-down development process of a complex system and discussing some aspects related to coordination and integration of development activities
- **Certification process and coordination,** discussing methods and techniques to demonstrate compliance of the system with certification requirements
- **Requirements determination and assignment of development assurance levels,** describing how the allocation of critical safety requirements determines the assurance levels of different system components
- **Safety assessment process,** providing an overview of the safety assessment activities and techniques. Notice that the safety assessment process is detailed in SAE (1996b)

- **Validation of requirements,** describing the process and the methods to demonstrate that the requirements are complete and correct for certification; the goal is to demonstrate that the requirements specify "the right system"
- **Implementation verification,** describing the process and the methods to demonstrate that the implementation implements all the requirements and only the requirements; the goal is to demonstrate that the implementation activity built "the system right"
- **Configuration management,** describing the activities that must be performed to ensure that all outputs of the development process form a coherent set of documentation
- **Process assurance,** describing the activities necessary to ensure that the development and support processes have been dutifully applied and executed

6.5.1 System Development

This part of the document describes how all the topics mentioned in the previous section integrate into a coherent development. In particular, the document presents a process model composed of two distinguished types of activities: those necessary for the development of a part of the system (development processes) and those necessary to support the overall development (supporting processes).

The development process is iterative, to capture increasing levels of refinement, both in the specification and in the implementation. Moreover, the development process "branches" together with the structure of a complex system. During each iteration, therefore, higher-level activities are performed just once, while more detailed activities might be instantiated various times to accommodate the independent development of different components and items. Finally, the development process runs parallel to, and integrates with, safety assessment activities.

The main system development activities envisaged by ARP4754 are

- **Aircraft-level functional requirements,** during which the high-level requirements and functions of the system are defined. (In parallel, safety assessment activities identify the critical functions.)
- **Allocation of aircraft functions to systems,** during which the functions identified at the previous step are allocated to systems. (In parallel, safety assessment activities determine assurance levels of systems.)
- **Development of system architecture,** during which the architecture is developed to implement the functions and satisfy the safety requirements individuated at the previous step. (In parallel, safety assessment activities determine assurance levels of components.)
- **Allocation of item requirements to hardware and software,** during which functions and safety requirements are allocated to items.
- **System implementation,** during which items are actually implemented.

Support processes run parallel to all the instantiations of the development process. Their goal is to provide an integrated and coherent view of all the activities characterizing the development of a complex system. The support processes are

- Certification coordination
- Safety assessment
- Requirements validation
- Implementation verification
- Configuration management
- Process assurance

6.5.2 Certification Process and Coordination

The document identifies three main steps for the certification:

1. **Certification planning,** which allows us to conciliate development constraints, system complexity, and certification activities.
2. **Agreement on the proposed means of compliance,** which allows the certification authority and the production team to agree on the means of compliance and to accommodate those special conditions that are necessary to properly take into account the specific characteristics of the system to be developed.
3. **Compliance substantiation,** that is, the actual implementation of the certification plan. During this activity, the agreed means of compliance are used to demonstrate that the system meets the airworthiness requirements.

Compliance substantiation is obtained through the production of data, that could include

- Aspects related to demonstrating that a principled development process has been followed (e.g., development, certification and validation plan, process assurance, configuration management)
- Analyses performed on the system (e.g., safety assessment, common cause analysis, system safety assessment)
- Aspects related to new technologies and their verification means

6.5.3 Assignment of Development Assurance Levels

One of the characterizing aspects of the document is the determination of development assurance levels, which define the rigor with which development must be performed on critical items. The document, in particular, identifies five different assurance levels (from A to E), which are determined based on the criticality of the function and are similar to that explained in Chapter 4.

Different criticality levels are associated to different architectural choices. In particular, the document distinguishes three different techniques to improve the safety of critical components and to control the effort necessary for system development:

- **Partitioning,** that is, the isolation to contain cross-functional effects when a component fails. Partitioning can be used to reduce the effort necessary for system development, by isolating the critical components from the noncritical ones and, consequently, by allowing for the use of lower levels of assurance for the noncritical components.
- **Redundancy,** that is, the use of redundant architectures (see Section 2.8).
- **Monitoring,** that is, the use of redundant architecture with hot or cold standby (see Section 2.8.5).

Furthermore, the document distinguishes among different types of redundant and monitoring architectures. In particular,

- **Similar designs,** if the redundant designs are based on the same physical phenomena and components.
- **Dissimilar independent designs,** if the redundant design are based on different physical phenomena and different components. Two braking systems, the first based on ABS on the wheels and the second based on a parachute that deploys on the ground, provide an example of dissimilar independent designs.
- **Dissimilar dependent designs,** if the designs are based on the same physical phenomena and different components. Two braking systems, the first based on ABS on the front wheels of a car and the second based on mechanical brakes on the rear wheels of the car, provide an example of dissimilar dependent designs.

Recognizing that redundant architectures help limit the degree to which the failure of a component can contribute to system failure, ARP4754 takes into consideration such reduced probability of error by granting lower levels of assurance, according to the chosen approach to the implementation of the redundant architecture. Thus, for instance, if a component is to be developed according to assurance level A, and if a dissimilar independent design is adopted, then the system establishing the independence must be developed according to Level A and the portions implementing the function in a dissimilar way can be implemented according to Level B.

Similarly, ARP4754 recognizes that the choice of a specific implementation technology or of the proper architecture can significantly mitigate or eliminate certain implementation errors. This, in turn, can make the achievement of certain objectives required by a specific assurance level more easily demonstrated or even readily demonstrated. Careful choices during implementation therefore allow us to control the complexity of system development.

Finally, the document considers the role that human intervention has in determining the assurance level. Although human intervention could, in principle, exacerbate failure conditions (and the Three Mile Island case study is an example of such a situation), if sufficient substantiation is provided, human intervention can be used to justify a lower assurance level for some components.

6.5.4 Safety Assessment

Concerning safety analysis, ARP4754 provides only information about what techniques must be used for the different assurance levels, as the detailed explanation of the safety assessment process is given in ARP4761.

In particular, Levels A to C require the achievement and demonstration that the probability of failure is below specific values given in the document. This is achieved through a controlled development process and by applying the safety analysis techniques that were described in Chapter 2 and Chapter 4.

6.5.5 Validation

ARP4754 intends with **validation** any activity meant to substantiate that the requirements are correct and complete. This amounts to demonstrating that there are no ambiguities, no incorrect statements, no incomplete statements, and no unintended functions. Notice, however, that because requirements live and exist at different level of detail (see, e.g., Chapter 4), validation in ARP4754 also deals with traceability.

Validation activities should be planned. According to the document, the validation process is composed of the following main steps:

- **Validation planning,** during which, also according to the assurance level, the methods to demonstrate compliance are determined.
- **Execution of checks,** during which the actual validation activities are performed. Different techniques can be used for validation. We mention analysis, modeling and testing, similarity, engineering judgment, and checklists. The applicable techniques depend on the assurance level. For instance, Levels A and B require analysis, modeling, and testing; Level C requires one of analysis, modeling, or testing.
- **Validation of assumptions,** during which all assumptions made during system development are actually listed and validated. Assumptions might include aspects such as the operational environment and other information that might not be available when system development begins. The techniques are similar to those applied for the validation of requirements.

Traceability among requirements plays an important role in ARP4754 because it constitutes essential information to guarantee, for example, that no unintended function is introduced during system development. Methods and techniques for traceability of requirements are up to the production team.

Parallel with the activities mentioned above, it is also necessary to maintain information about the traceability of requirements and results.

6.5.6 Implementation Verification

This process has the goal of ensuring that the system implementation satisfies the validated requirements. The main goal is to demonstrate that the system performs all the functions and only the functions for which it has been devised. Moreover, it serves to demonstrate that the results of preliminary safety analysis are still valid for the system as implemented.

The implementation verification process uses different techniques—that depend also on the assurance level—among which are

- **Inspections and reviews,** typically based on checklists, are meant to demonstrate that the implementation complies with the requirements.
- **Analysis** provides evidence of compliance through detailed examination. Analysis can be based on modeling and coverage analysis. The goal is to demonstrate coverage of requirements.
- **Testing** is used to demonstrate implementation of requirements and, when possible, to demonstrate that the implemented system does not perform unintended functions that could impact safety.
- **Service experience** can be used whenever a component that has been already certified is reused in a new design, according to its specification.

6.6 ARP 4761

ARP4761, "Guidelines and Methods for Conducting the Safety Assessment Process on Civil Airborne Systems and Equipment," describes a set of guidelines for performing safety assessment for the certification of civil aircraft. The document is organized in two main parts—"Safety Assessment Process" and "Safety Assessment Analysis Methods"—describing, respectively, a process for a systematic assessment of a complex system and the techniques that can be used to support the execution of the process.

The first part of the document identifies the safety analysis process, which, similar to that described in Chapter 4, consists of the following main activities:

- **Fault Hazard Analysis (FHA).** During this activity, all failures of the system are classified according to their severity. A justification for the classification is provided. This allows us to determine target probabilities of occurrence for each different failure. The assessment can be performed at two different levels of granularity, aircraft and system, and initial failure modes per function can be decomposed into failure modes of subsystems and components.

◼ **Preliminary System Safety Assessment (PSSA).** During this activity, all the failures identified in the previous step are allocated to design components, in order to achieve the desired/necessary safety conditions. PSSA defines the safety requirements for components, such as, for example, the maximum tolerable failure rate.

◼ **System Safety Assessment (SSA).** During this activity, an actual design is evaluated with respect to the target goals of the safety analysis. The activity differs from PSSA in the sense that it is a verification activity, rather than a way of determining the requirements of the different components. The starting point is, once again, the failures identified in the fault hazard analysis. During this phase, techniques such FTA, FMEA, and FMES are used to analyze different components and evaluate whether or not they meet the target goals of the evaluation.

As already explained in Chapter 4, it is essential to conduct **Common Cause Analysis** during SSA, to ensure that the independence hypothesis holds and that all common failure conditions are properly identified and dealt with.

The second part of the document describes the techniques that can be used to support safety assessment. Most of them were described in Chapter 3. We mention Fault Tree Analysis, Dependency Diagrams, Markov Analysis, Failure Mode and Effect Analysis, and Common Cause Analysis.

6.7 DO-178B

This document, prepared jointly by two groups, EUROCAE WG 12 and RTCA Special Committee 167 and published in December 1992, is meant to satisfy the need for common guidance in the development of software systems that could satisfy airworthiness requirements. It is a mature and complete document that collects over 20 years of best practices and consensus in industry. Although the document specifically mentions the aeronautics as its target, the document is easily applicable to other (safety-) critical domains. See Spitzer (2006) for more information on the impact and scope of this document.

6.7.1 Document Overview

Different from ARP4754 and from other standardization documents, DO-178B (ED-12B) does not endorse or propose a specific development process or a specific software development methodology. (Remark: in the remainder of the section we indifferently use ED-12B and DO-178B to refer to the standard.) Quoting:

"ED-12B recognizes that many different software life cycles are acceptable for developing software for airborne systems and equipment. ED-12B emphasizes that the chosen software life cycle(s) should be defined during the planning for a project."

Certification under DO-178B is, instead, achieved by demonstrating compliance of the development process and of the artifact produced during development with a set of goals. The goals depend on three factors:

■ Development activity type
■ Software category
■ Control category

Development Activity Type. Although DO-178B does not endorse any specific process, it is rather clear that a disciplined approach is needed to achieve certification. Goals in the document are thus specified for the three different types of activity that characterize system development; they are

■ **Software planning process,** which defines and organizes development and support activities.
■ **Development process,** which has the goal of building a software product.
■ **Integral processes,** which support and complement the development process by ensuring the correctness and quality of the final system, and control over system configuration. The integral processes include **software verification**, **software configuration management**, **software quality assurance**, and **certification liaison**.

Software Category. Similar to ARP4754 and ARP4761, DO-178B distinguishes software systems in five classes, from A to E, according to the effect of a software malfunction.

Control Category. The document introduces two different control categories —CC1 and CC2—corresponding to two different policies for configuration control and for managing the artifacts of the process. CC1 and CC2 are defined through thirteen characteristics, some of which are technical and some of which are organizational. We mention, for instance, *Protection against unauthorized changes,* which must be achieved in both control categories.

The control categories are introduced to recognize different costs and trade-offs in software development. Of the two, CC1 is more stringent and it must be applied for the most critical outputs of the process and for the more critical software categories.

6.7.2 Goals

For each goal, the document provides the following information:

- **Description of the goal** in natural language. For instance, "High-level requirements are developed."
- **Applicability,** that allocates the achievement of the objective to one or more software category. Notice that, for any given software category, there are three different possible values:
  - The goal does not need to be achieved.
  - The goal must be achieved.
  - The goal must be achieved with independence, when a separation of responsibilities between accomplishment of objective and verification is required.
- **Output,** which is the artifact obtained by achieving the goal.
- **Control category,** in which the control category of the artifact is specified according to the software category (one of CC1 or CC2).

Goals might vary quite a bit in scope and difficulty. The document defines objectives that are relatively easy to achieve—for example, "Software life cycle environment is defined"—and goals whose achievement is rather more difficult to demonstrate—"Algorithms are accurate."

Thus, although the effort to achieve each goal is rather different, summarizing the number of goals per type of activity is a useful exercise that allows us to understand where the standard places the most emphasis. This is shown in Table 6.2, where the rows indicate the different software processes and the columns indicate the software levels and the objectives to achieve, distinguished among those that must be achieved with independence from those that do not require independence. (See also Spitzer (2006) for a similar discussion.)

DO-178B places a lot of emphasis on testing activities. This can be easily shown by looking at plots in Figure 6.5, which graphs the number of goals per level. From the diagram we can easily see that

- Most of the goals are relative to the verification processes. For instance, for the development of Level A software, they are four times (40) those defined for the other processes (about 10 per process). This is partly due to the scope of goals. For instance, there is just one development goal related to coding, namely "Source code is developed" and there are seven goals related to verifying the source code. This is, however, also a demonstration of the greater analytical work and detail that the document puts in verification activities.
- As we move toward less critical software (i.e., from Level A to Level D), goals related to planning and to testing are decreasing. This indirectly indicates the role that planning and verification activities have in increasing software quality.

Table 6.2 The Goals of DO-178B

	A			B			C			D		
	Required with Indep.	Required	Total for Level A	Required with Indep.	Required	Total for Level B	Required with Indep.	Required	Total for Level C	Required with Indep.	Required	Total for Level D
Software Planning		7	7		7	7		7	7		2	2
Software Development		7	7		7	7		7	7		7	7
Verification of Requirements	3	4	7	3	4	7		6	6		3	3
Verification of Design	6	7	13	3	10	13		9	9		1	1
Verification of Coding	3	4	7	1	6	7		6	6			0
Integration Testing	2	3	5	1	4	5		5	5		3	3
Verification of Process	8		8	3	5	8		5	5		1	1
Configuration Management		6	6		6	6		6	6		5	5
Software Quality Assurance	3		3	3		3	2		2	2		2
Certification		3	3		3	3		3	3		3	3
Total	25	41	66	14	52	66	2	54	56	2	25	27

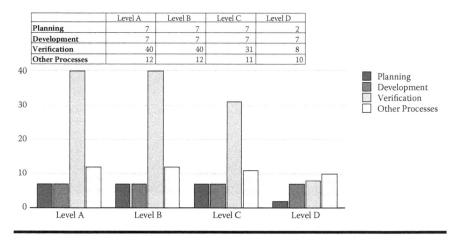

	Level A	Level B	Level C	Level D
Planning	7	7	7	2
Development	7	7	7	7
Verification	40	40	31	8
Other Processes	12	12	11	10

Figure 6.5   Some data from the DO-178B goals.

6.7.3 Verification Activities

Verification is defined as a combination of reviews, analyses, and testing:

- **Reviews** are typically conducted on the higher-level artifacts (e.g., requirements and design) and are conducted using the common practices defined in software engineering, for example, by using checklists and looking for inconsistencies, ambiguities, and incompleteness. See, for instance, Sommerville (2007) for a discussion on the matter and NASA (2009) for a concrete example.
- **Analyses** are repeatable in nature and, therefore, are often algorithmic or procedural in nature (Spitzer, 2006). They might include coverage analysis, data, and control flow analyses. See Sommerville (2007) for a detailed description of the analyses.
- **Testing** is requirements based. The practice, quite standard now in software development, was a great shift with respect to the first version of the document (Spitzer, 2006). As mentioned earlier, the number of goals related to testing changes according to the software level. We mention here that for all levels, the traceability of test cases to high-level requirements must be demonstrated, whereas the traceability of test cases to low-level requirements is required for Levels A and B only.

For Level A and B, test cases must also demonstrate different types of code coverage, including

- Decision coverage for Levels A and B.
- Statement coverage for levels A, B, and C.
- Modified condition/decision for Level A. This is more stringent than decision coverage, because, intuitively, it requires one to verify both the true and false condition for any leaves of any complex Boolean expression. For instance, if

the condition is ($A \wedge B$), all possible values of A and B must be tested. See Hayhurst et al. (2001) for more details.

■ Data coupling and control coupling for Level A. The goal of this activity is "to provide a measurement and assurance of the correctness of these modules/components, interactions and dependencies. That is, the intent is to show that the software modules/components affect one another in the ways in which the software designer intended and do not affect one another in ways in which they were not intended, thus resulting in unplanned, anomalous, or erroneous behavior." (Certification Authorities Software Team, 2004).

Notice also that all goals also require independent achievement for Level A software, while only the coverage goals require independent achievement for Level B.

6.7.4 Certification Goals

Certification aspects are described in Section 10 of DO-178B, where three goals are set:

1. **Certification basis,** that is, an agreement between the certification authority and the applicant over the regulations and special conditions that apply to the system to be certified.
2. **Assessment of the "Plan for Software Aspects of Certification,"** that is, the document that describes how the applicant intends to achieve and demonstrate compliance with the regulations. The plan is analyzed for "completeness and consistency with the means of compliance" and "with the outputs of the system safety assessment process" (in order to ensure coherence between the analyses performed at the different levels of granularity of the system).
3. **Compliance of the software,** finally, is assessed by analyzing the "Software Accomplishment Summary," which includes all the relevant information related to the software system, among which are
   ■ An overview of the system and of the software, together with certification considerations and the main characteristics of the system (e.g., size, memory margins, software identification)
   ■ Software life cycle and data produced during the project
   ■ Software history, such as change history and unresolved issues at the time of certification

(See EUROCAE (1992) for a more detailed description of each item.)

6.7.5 The Role of Tools

Under DO-178B, tools can be used to support and simplify the development, assessment, and certification processes. The use of tools to support the achievement of one or more goals is rather simple: any tool can be used as long as its outputs are then verified through some other means. That is, the tool can be used to automate certain tasks but cannot be a substitute for verification activities.

Under special conditions, however, tools can be used to demonstrate the achievement of a goal. This gives us the opportunity to significantly simplify the certification process.

DO-178B distinguishes between two types of tools:

- **Software development tools:** tools whose output is part of airborne software
- **Software verification tools:** tools that cannot introduce errors but that might fail to detect them

The qualification of software development tools is rather articulated and requires one to demonstrate that two conditions are met. The first is that the tool has been developed meeting the same objectives and the same assurance level for which it is applied. The second is to demonstrate that the applicant has used the tool under the conditions for which it was designed (EUROCAE, 1992).

By contrast, qualification of software verification tools is simpler because it requires the applicant to demonstrate that the tools have been used according to their operational requirements.

The advantages that can be obtained by introducing tools in the development and certification process are rather clear. As a result, various vendors have certified their tools under DO-178B or provide guidelines on how to demonstrate compliance with their tools' operational requirements.

6.7.6 The Role of Formal Methods in DO-178B

DO-178B has a specific part dedicated to describing the role of formal methods for certification. This is in the "Alternative Methods" chapter because of "their inadequate maturity at the time this document [DO-178B] was written." However, formal methods can be used as long as they are demonstrated to address the goals of the standard, and their usage is adequately planned and described in the "Plan for Software Aspects of Certification."

6.8 The Case for the Safety Case

The safety case is an approach complementary to demonstrating compliance with regulatory norms and meant to show that a system can be safely operated. Using the definition given in Wilson et al. (1997),

> "The purpose of a safety case is to present a clear, comprehensive and defensible argument supported by calculation and procedure that a system or installation will be acceptably safe throughout its life (and decommissioning)."

There are two important motivations behind the introduction of the safety case. The first is focus: Safety cases are meant to demonstrate the actual safety of a system, rather than compliance of the system with prescriptive regulations. This helps ground safety analyses, which do not rely anymore on the hypotheses that the prescriptive regulations are sufficient to guarantee the safety of a system. The second concerns responsibility, which is shifted back to the production team. This is, in fact, responsible for finding convincing arguments to demonstrate system safety. Safety regulations can be such arguments, but only as long as they are critically evaluated and their applicability to the safety case sufficiently motivated.

In John and McDermid (2001) we find the three main elements that constitute a safety case:

- **Safety requirements and objectives,** which define what are the goals of the analysis
- **Safety evidence,** which defines what the evidences are on which the analyses rely
- **Safety argument,** which describes and argues how the safety evidence is sufficient to demonstrate the achievement of the safety objectives

Notice that both the evidence and the argument are necessary because, as explained in Kelly (2003): "Argument without supporting evidence is unfounded and therefore unconvincing. Evidence without argument is unexplained—it can be unclear that (or how) safety objectives have been satisfied."

The safety argument can use different approaches, such as descriptions in natural language, formal inference rules, and the techniques that we have presented in this book. Often, however, it can be presented as a structured argumentation, similar to a mathematical proof, or a tree such as that shown in Figure 6.6.

See also Papadopoulos and McDermid (1999) for some considerations related to the construction of safety cases, and EUROCONTROL (2006), for a methodology to build safety cases.

Finally, we mention that safety cases can evolve and can be progressively refined as the system develops. For instance, the Ministry of Defence (1997) distinguishes among the following:

- **Preliminary safety case,** produced after the system requirements have been defined. The goal is to justify the way in which the Software Safety Plan is expected to deliver System Requirement Specifications that meet the safety requirements of the specified safety integrity level.
- **Interim safety case,** produced after the specification and is meant to demonstrate that the safety requirements are met by the specification.
- **Operational safety case,** which includes the set of evidence that the safety requirements have been met for the specified safety integrity level.

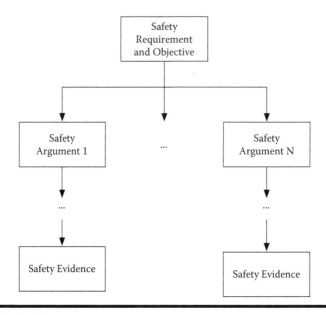

Figure 6.6   A safety case.

6.9 Formal Methods and Certification

As shown in Chapter 5, formal methods have the potential to efficiently support development, assessment, and validation of complex safety-critical systems. Moving formal methods out of the laboratory, however, has proven difficult, as also highlighted by the DO-178B standard. This is due to various issues and points of attention in the usage of formal techniques, which were discussed in Section 5.4. When we consider the certification of complex systems, aspects such as training, tool robustness, and tool qualification can certainly be added to the list of issues that should be addressed for an effective integration of formal methods.

We would like to conclude this book by briefly describing an approach to the integration of formal methods in the development process. The example is interesting, in our opinion, because it shows how a proper mix of techniques and methodologies allows one to define new paradigms that have the potential to significantly change the approach to system development, assessment, and certification.

Understanding how formal methods can be integrated into the development process is a topic that is more than 20 years old. We cite Kemmerer (1990), which describes three different approaches for software systems:

1. **After the fact,** when the system is built using standard techniques and, only after it has been completed, a formal specification for the system is written and properties are proved about the specification.

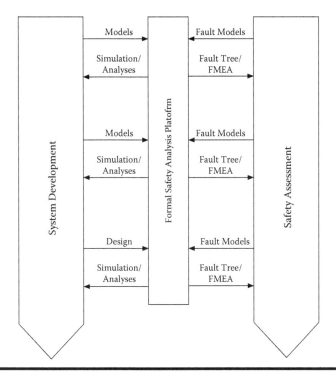

Figure 6.7   Integrating formal methods in the development of complex systems.

2. **Parallel,** when formal specification activities are performed in parallel to system development activities. Development activities are carried out using standard techniques. These, however, are used to come out with formal specifications, which are verified in parallel with the development activities.
3. **Integrated,** when formal specifications are used as the main techniques to drive system development activities.

When we focus on the development of complex safety-critical applications, of the three approaches, the parallel application of formal methods seems most promising. This is due to the fact that emerging methodologies such as ESACS and ISAAC, described in Section 5.8, promote and encourage integration between design and safety assessment. This is shown in Figure 6.7, where on the left-hand side we have the system development activities, on the right-hand side we have the safety assessment activities, and in the middle we have tools supporting the ESACS and ISAAC methodology. As development activities progress, system designers build formal models, at increasing levels of detail, that they can use to perform simulation and analyses. These models are also used by Safety Analysts, who can automatically "extend" them with failure modes and, thus, assess the safety of the design.

Of course, as system complexity increases, we expect that formal methods will be less effective, due to the increasing complexity of systems and the state explosion problem. Among the advantages, however, are that the use of a common language and infrastructure for system specification and safety assessment allows both system designers and safety engineers to speak the same unambiguous language and share the same formal system model. In addition, the safety evaluation of the system architecture can be performed in the very early phases of system design, by simulating and proving properties of the system model.

References

AIA, GAMA, and FAA Aircraft Certification Service (2004). The FAA and Industry Guide to Product Certification. Available at `http://www.faa.gov/aircraft/air-cert/design_approvals/media/CPI_guide_II.pdf`. Last retrieved on November 15, 2009.

Certification Authorities Software Team (2004). Clarification of Structure Coverage Analyses of Data Coupling and Control Coupling. Position Paper CAST-19, Federal Aviation Administration. Last retrieved on November 15, 2009.

Department of Defense (2005). Airworthiness Certification Criteria. Technical Report MIL-HDBK-516B, Department of Defense.

Department of Defense (Last retrieved on November 15, 2009). Assist. Available at `https://assist.daps.dla.mil/online/start/`.

ECSS Secreteriat, ESA-ESTEC (2008). ECSS system—Description, Implementation, and General Requirements. ST ECSS-S-ST-00C, European Space Agency.

EUROCAE (1992). Software Considerations in Airborne Systems and Equipment Certification (DO-178B/ED-12B). Technical Report DO-178B/ED-12B, EUROCAE.

EUROCAE (Last retrieved on November 15, 2009). EUROCAE web site: `http://www.eurocae.net`.

EUROCONTROL (2006). Safety Case Development Manual. Technical Document DAP/SSH/091, European Union.

European Cooperation for Space Standardization (Last retrieved on November 15, 2009). European Cooperation for Space Standardization web site. Available at `http://www.ecss.nl/`.

Federal Aviation Administration (1993). RTCA/DO-178B, Software Considerations in Airborne Systems and Equipment Certification. AC 20-115B, Federal Aviation Administration. Last retrieved on November 15, 2009.

Federal Aviation Administration (1998). System Design and Analysis. Advisory Circular 25.1309-1A, U.S. Deparment of Transportation.

Federal Aviation Administration (2005). RTCA/DO-254, Design Assurance Guidance for Airborne Electronic Hardware. AC 20-152, Federal Aviation Administration. Last retrieved on November 15, 2009.

Federal Aviation Administration (2007). Type Certification. Order 8110.4C, U.S. Deparment of Transportation.

Hayhurst, K.J., D.S. Veerhusen, J.J. Chilenski, and L.K. Rierson (2001). A Practical Tutorial on Modified Condition/Decision Coverage. Technical Report TM-2001-210876, NASA.

John, T.K. and J. McDermid (2001). A systematic approach to safety case maintenance. In M. Felici, K. Kanoun, and A. Pasquini (Eds.), *Proc. 18th International Conference on Computer Safety, Reliability and Security (SAFECOMP'99)*, Volume 1968 of *LNCS*, pp. 13–26. Springer.

Kelly, T. (2003). A Systematic Approach to Safety Case Management. Technical Report 04AE-149, SAE.

Kemmerer, R.A. (1990). Integrating formal methods into the development process. *IEEE Software* 7(5), 37–50.

Ministry of Defence (1997). Requirements for Safety Related Software in Defence Equipment. Technical Report 00-55, Ministry of Defence.

Ministry of Defence (Last retrieved on November 15, 2009). UK Defence Standardization web site. Available at `http://www.dstan.mod.uk/`.

NASA (Last retrieved on November 15, 2009). Software Requirements Review (SRR) Checklist. Available at `http://sw-assurance.gsfc.nasa.gov/disciplines/quality/checklists/pdf/software_requirements_review.pdf`.

Papadopoulos, Y. and J.A. McDermid (1999). The potential for a generic approach to certification of safety-critical systems in the transportation sector. *Journal of Reliability Engineering and System Safety* 63(47–66).

RTCA (2000). Design Assurance Guidance for Airborne Electronic Hardware. Technical Report DO-254. Washington, D.C.: RTCA, Inc.

SAE (1996a). Certification Considerations for Highly-Integrated or Complex Aircraft Systems. Technical Report ARP4754, Society of Automotive Engineers.

SAE (1996b). Guidelines and Methods for Conducting the Safety Assessment Process on Civil Airborne Systems and Equipment. Technical Report ARP4761. Warrendale, PA: Society of Automotive Engineers.

SAE (Last retrieved on November 15, 2009). Society for Automotive Engineers home page. `http://www.sae.org`.

Sheard, S.A. (2001). Evolution of the framework's quagmire. *Computer* 34(7), 96–98.

Sommerville, I. (2007). *Software engineering (8th ed.)*. Reading, MA: Addison-Wesley.

Spitzer, C.R. (Ed.) (2006). *Digital Avionics Handbook—Avionics, Elements, Software, and Functions* (2nd ed.). Boca Raton, FL: CRC.

U.S. Government (Last retrieved on November 15, 2009). Code of Federal Regulations, Title 14—Aeronautics and Space. Available at `http://ecfr.gpoaccess.gov/cgi/t/text/text-idx?c=ecfr&tpl=/ecfrbrowse/Title14/14tab_02.tpl`.

Wilson, S.P., T. Kelly, J.A. McDermid, and Y. England (1997). Safety case development: Current practice, future prospects. In *12th Annual CSR Workshop on Safety and Reliability of Software Based Systems*. Berlin: Springer.

Appendix A

The NuSMV Model Checker

In this appendix we give a short overview of the NuSMV (NuSMV, 2009; Cimatti et al., 2000, 2002) model checker and its main features.

A.1 Introduction

NuSMV is a symbolic model checker (see Section 5.7.3), originally developed as a joint project between the Institute for the Scientific and Technological Research of Fondazione Bruno Kessler (FBK, 2009) (formerly Istituto Trentino di Cultura), Carnegie Mellon University, University of Trento, and University of Genova. It originates from a reimplementation, reengineering, and extension of SMV (McMillan, 1993), the first symbolic model checker, based on BDDs (Bryant, 1992). NuSMV extends SMV along several directions, namely new functionalities, new architecture, and implementation.

NuSMV supports state-of-the art model checking technologies, such as BDD- and SAT-based verification for CTL and LTL temporal logics (see Section 5.6.6). NuSMV is a well-structured, flexible, customizable, robust, and documented platform for model checking. It was designed as an open architecture, which can be reliably used for the verification of industrial designs, as a core for custom verification tools, as a testbed for formal verification techniques, and can be applied to several research areas.

A.2 Distribution

Starting with version 2, NuSMV is released with an OpenSource license (OS, 2009), namely the GNU Lesser General Public License (LGPL) (LGPL, 2009). The open source schema allows anyone interested to freely use the tool and contribute to its development. Furthermore, it provides the model checking community with a common platform for research, and implementation and experimental evaluation of new model checking techniques.

The open source schema also allows for evaluation and independent peer review of the system and its implementation, resulting in faster system evolution and higher-quality implementation. Since its first distribution in open source, NuSMV has received several contributions for different parts of the system, and has attracted interest from many groups for both use and development.

The LGPL license grants NuSMV users full rights to use and modify the program, for both research and commercial applications, either as a stand-alone tool or within bigger software systems, with the only proviso that any improvement and extension to the original system should be made available under the same license.

A.3 Architecture

NuSMV is designed with a highly modular and open architectural scheme, greatly revised starting with version 2. The architecture of the system was conceived with the goal of making most of the functionalities of the tool independent of the specific model checking algorithms used.

The architecture of the system is structured in several modules that implement specific functionalities and communicate with each other via well-defined software interfaces. This modular and open architecture allows easy removal, addition, or replacement of any existing module.

A.4 Implementation

The NuSMV system is highly portable, maintainable, and robust. It is written in ANSI C and is POSIX compliant. NuSMV has been successfully compiled on several operating systems. The source code is structured in several packages and maintained using tools for version control. Moreover, add-ons are available for specific applications, for example, an addon implementing safety analysis routines used by the FSAP tool (see Appendix B).

The BDD-based model checking routines exploit the CUDD library (CUDD, 2009) for BDDs developed by Fabio Somenzi at Colorado University. Furthermore, NuSMV provides links with external solvers for SAT-based Bounded Model Checking verification, such as zChaff (zChaff, 2009) and MiniSat (MiniSat, 2009).

A.5 User Interaction

NuSMV has both a batch and an interactive mode of execution. In batch mode, verification routines are activated on the basis of a predefined algorithm. In interactive mode, user commands are executed by a command-line interpreter, thus allowing for the decomposition and customization of the predefined model checking algorithms. Using the interactive shell, compilation of a model is decomposed in several steps (parsing, flattening of the hierarchy, encoding, compilation into BDDs). The interactive shell allows the user to invoke commands separately, possibly repeating or undoing some of them. Moreover, it provides the possibility to fully configure the runtime options of the model checker, including setting several options of the BDD package.

A.6 Language

The NuSMV language provides several data types, including Boolean variables, bounded integer subranges, sets, symbolic enumerations, words, and bounded arrays of basic data types.

A NuSMV program is structured in modules, which can be instantiated several times. As discussed in Section 5.6.5, NuSMV supports both synchronous and asynchronous composition. Using synchronous composition, a single step of the composition corresponds to a single step in each on the modules. Asynchronous composition uses interleaving; that is, each step of the composition corresponds to a single step performed by exactly one component. The NuSMV language allows us to describe both deterministic and nondeterministic systems.

Properties can be written in either CTL, LTL, or PSL temporal logic (see Section 5.6.6). Some examples of NuSMV programs are given in Section 5.6.5. For more details and a complete discussion of the syntax and semantics of NuSMV, and examples of its use, refer to the NuSMV user manual (Cavada et al., 2009a) and the NuSMV tutorial (Cavada et al., 2009b). Several test suites and examples are distributed together with the NuSMV source code.

A.7 Functionalities

NuSMV provides several functionalities, including BDD-based CTL model checking, BDD-based and SAT-based LTL model checking, reachability analysis, invariant checking, and quantitative characteristics computation. An important feature of model checking is the possibility of generating counterexample traces whenever a specification is violated.

NuSMV implements state-of-the-art model checking technologies, such as dynamic reordering of BDD variables (inherited from the CUDD package), cone of influence reduction (Berezin et al., 1998), and several strategies for conjunctive partitioning (Clarke et al., 2000).

A.8 Applications

NuSMV is being used for educational and teaching purposes, and has been used as a basis for several Ph.D. theses. Furthermore, over the years it has attracted interest from several industrial companies.

Moreover, NuSMV has been used for technology transfer in several industrial projects and applications, including formal validation of requirements (Fuxman et al., 2004; Pill et al., 2006; Cavada et al., 2009c), planning (Bertoli et al., 2001; Cimatti et al., 2003b) diagnosability (Cimatti et al., 2003a; Bertoli et al., 2007), and aerospace applications (Bozzano et al., 2008a,b, 2009a,b). Several tools have been built on top of NuSMV; among these we mention

- The MBP Model Based Planner (MBP, 2009)
- The RAT Requirements Analysis Tool (RAT, 2009)
- The FSAP Safety Analysis Platform (FSAP, 2009)

The FSAP platform and applications to safety analysis are described in more detail in Appendix B.

References

Berezin, S., S. Campos, and E.M. Clarke (1998). Compositional reasoning in model checking. In W.P. de Roever, H. Langmaack, and A. Pnueli (Eds.), *Proc. International Symposium on Compositionality: The Significant Difference (COMPOS'97)*, Volume 1536 of *LNCS*, pp. 81–102. Berlin: Springer.

Bertoli, P., M. Bozzano, and A. Cimatti (2007). A symbolic model checking framework for safety analysis, diagnosis, and synthesis. In *Model Checking and Artificial Intelligence*, Volume 4428 of *LNCS*, pp. 1–18. Berlin: Springer.

Bertoli, P., A. Cimatti, M. Pistore, M. Roveri, and P. Traverso (2001). MBP: A model based planner. In *Proc. Workshop on Planning under Uncertainty and Incomplete Information*.

Bozzano, M., G. Burte, A. Cimatti et al. (2008a). System and Software Co-Engineering: Performance and Verification. *ESA Workshop on Avionics Data, Control and Software Systems (ADCSS)*.

Bozzano, M., A. Cimatti, and A. Guiotto et al. (2008b). On-Board Autonomy via Symbolic Model Based Reasoning. *10th ESA Workshop on Advanced Space Technologies for Robotics and Automation (ASTRA 2008)*.

Bozzano, M., A. Cimatti, J.-P. Katoen, V.Y. Nguyen, T. Noll, and M. Roveri (2009a). The COMPASS approach: correctness, modelling and performability of aerospace systems. In B. Buth, G. Rabe, and T. Seyfarth (Eds.). In *Proc. 28th International Conference on Computer Safety, Reliability and Security (SAFECOMP 2009)*, Volume 5775 of *LNCS*, pp. 173–186. Berlin: Springer.

Bozzano, M., A. Cimatti, M. Roveri, and A. Tchaltsev (2009b). A comprehensive approach to on-board autonomy verification and validation. In *Proc. ICAPS'09 Workshop on Verification and Validation of Planning and Scheduling Systems (VV&PS 2009)*.

Bryant, R.E. (1992). Symbolic Boolean manipulation with ordered binary decision diagrams. *ACM Computing Surveys* 24(3), 293–318.

Cavada, R., A. Cimatti, and C.A. Jochim et al. (Last retrieved on November 15, 2009a). NuSMV 2.4 User Manual. Available at `http://nusmv.fbk.eu/NuSMV/userman/v24/nusmv.pdf`.

Cavada, R., A. Cimatti, G. Keighren, E. Olivetti, M. Pistore, and M. Roveri (Last retrieved on November 15, 2009b). NuSMV 2.2 Tutorial. Available at `http://nusmv.fbk.eu/NuSMV/tutorial/v24/tutorial.pdf`.

Cavada, R., A. Cimatti, A. Mariotti et al. (2009c). EuRailCheck: Tool support for requirements validation. In *Proc. 24th IEEE/ACM International Conference on Automated Software Engineering (ASE 2009)*. Washington, D.C.: IEEE Computer Society.

Cimatti, A., E.M. Clarke, F. Giunchiglia, and M. Roveri (2000). NuSMV: a new symbolic model checker. *Software Tools for Technology Transfer* 2(4), 410–425.

Cimatti, A., E.M. Clarke, E. Giunchiglia et al. (2002). NuSMV2: An opensource tool for symbolic model checking. In E. Brinksma and K. Larsen (Eds.), *Proc. 14th International Conference on Computer Aided Verification (CAV'02)*, Volume 2404 of *LNCS*, pp. 359–364. Springer.

Cimatti, A., C. Pecheur, and R. Cavada (2003a). Formal verification of diagnosability via symbolic model checking. In G. Gottlob and T. Walsh (Eds.), *Proc. 18th International Joint Conference on Artificial Intelligence (IJCAI 2003)*, pp. 363–369. San Francisco, CA: Morgan Kaufmann.

Cimatti, A., M. Pistore, M. Roveri, and P. Traverso (2003b). Weak, strong, and strong cyclic planning via symbolic model checking. *Artificial Intelligence* 147(1-2), 35–84.

Clarke, E.M., O. Grumberg, and D.A. Peled (2000). *Model Checking*. Cambridge, MA: MIT Press.

CUDD (Last retrieved on November 15, 2009). CUDD: CU Decision Diagram Package. `http://vlsi.colorado.edu/~fabio/CUDD`.

FBK (Last retrieved on November 15, 2009). Fondazione Bruno Kessler. `http://www.fbk.eu`.

FSAP (Last retrieved on November 15, 2009). The FSAP/NuSMV-SA platform. `https://es.fbk.eu/tools/FSAP`.

Fuxman, A., L. Liu, J. Mylopoulos, M. Pistore, M. Roveri, and P. Traverso (2004). Specifying and analyzing early requirements in Tropos. *Requirements Engineering* 9, 132–150.

LGPL (Last retrieved on November 15, 2009). The GNU Lesser General Public License. `http://www.fsf.org/licensing/licenses/lgpl.html`.

MBP (Last retrieved on November 15, 2009). The MBP Model Based Planner. `http://mbp.fbk.eu`.

McMillan, K.L. (1993). *Symbolic Model Checking*. Dordrecht, The Netherlands: Kluwer Academic.

MiniSat (Last retrieved on November 15, 2009). The MiniSat Page. `http://minisat.se`.

NuSMV (Last retrieved on November 15, 2009). NuSMV: A New Symbolic Model Checker. `http://nusmv.fbk.eu`.

OS (Last retrieved on November 15, 2009). The Open Source Initiative. `http://www.opensource.org`.

Pill, I., S. Semprini, R. Cavada, M. Roveri, R. Bloem, and A. Cimatti (2006). Formal analysis of hardware requirements. In E. Sentovich (Ed.), *Proc. 43rd Design Automation Conference (DAC'06)*, pp. 821–826. New York: ACM.

RAT (Last retrieved on November 15, 2009). The RAT Requirements Analysis tool. `http://rat.fbk.eu`.

zChaff (Last retrieved on November 15, 2009). ZChaff. `http://www.princeton.edu/~chaff/zchaff.html`.

Appendix B

The FSAP Safety Analysis Platform

In this Appendix we give a short overview of the FSAP (Bozzano and Villafiorita, 2007; FSAP, 2009) safety analysis platform and its main features.

B.1 Introduction

The Formal Safety Analysis Platform (FSAP) aims to support the development, formal analysis, design, and safety assessment of complex systems. The platform implements the so-called ESACS methodology, described in Section 5.8, for model-based safety analysis.

As one of its main characteristics, the platform provides a facility for automatically augmenting a system model with failure modes, whose definitions are retrieved from a predefined library. The addition of failure mode definitions to a system model is called *fault injection* (compare Section 5.8.1), and the related extension step is called *model extension* (compare Section 5.8.2). Using the mechanism of model extension, it is possible to assess system safety both in nominal conditions and in user-specified degraded situations, that is, in the presence of faults. In this way, the effort is placed on the design and safety engineers can focus on building formal models of the system, rather than carrying out the analyses.

The FSAP platform is based on the NuSMV model checker (described in Appendix A), which provides its formal verification capabilities. Moreover, it provides a graphical user interface. The platform has been developed and is copyrighted by the Institute for the Scientific and Technological Research of Fondazione Bruno Kessler (formerly Istituto Trentino di Cultura) (FBK, 2009). FSAP has been developed within the ESACS (Enhanced Safety Assessment for Complex Systems) (ESACS, 2009) and ISAAC (Improvement of Safety Activities on Aeronautical Complex systems) (ISAAC, 2009) projects, two European-Union-sponsored projects involving various research centers and industries in the fields of avionics and aerospace.

FSAP is designed to support different phases of the development and safety assessment of complex systems. Toward this aim, FSAP provides a set of basic functionalities that can be combined to perform complex tasks. The main benefits of the platform are a tighter integration between design and safety assessment, based on the use of a shared formal language, and the automation of some of the activities related both to system verification and safety assessment, in a uniform environment. Moreover, the functionalities provided by the platform can be combined in different ways, enabling the integration of the platform in different development and safety analysis processes (e.g., based on incremental development).

B.2 Distribution

The FSAP platform is currently distributed for evaluation and teaching purposes, upon signing a bilateral agreement between Fondazione Bruno Kessler and the interested party. Moreover, licensing for commercial applications is under consideration (see FSAP2, 2009). Inquiries for licensing can be addressed to the mailing list fsap@fbk.eu.

B.3 Architecture

Architecturally, FSAP is based on two main components:

1. A graphical user interface, which provides access to all the functionalities of the platform, including modeling, verification and traceability facilities, and results displayers[1].
2. A verification engine, called NuSMV-SA, which is a NuSMV add-on (see Appendix A), and implements the core algorithms for formal verification.

The architecture of the FSAP platform is summarized in Figure B.1.

The different blocks comprising the architecture are described in more detail below.

■ **SAT Repository.** The *Safety Analysis Task* (SAT) Repository is the central module of the platform. It is used to store all the information relevant to modeling, verification, and safety assessment. In particular, it contains references to the (nominal and extended) system models (see Section 5.8), the definition of the

---

[1] The basic functionalities of the platform can also be accessed from the command line, using a batch interaction.

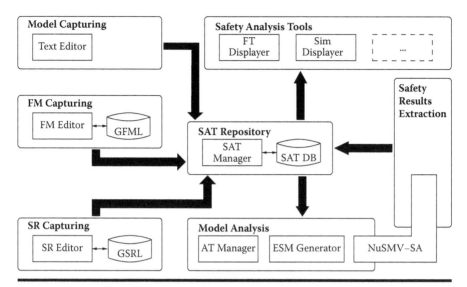

Figure B.1   The architecture of the FSAP platform.

failure modes and safety requirements, and analysis tasks and the corresponding results. The SAT Repository is composed of an SAT Database (SAT-DB) containing all sorts of information, and an SAT Manager, which retrieves the relevant information from the SAT Database. All the other components of the platform are accessed through the SAT repository. The information pertaining to a specific project is saved in XML format and can be stored and retrieved from the file system.

■ **Model Capturing.** The Model Capturing block is responsible for managing input models. As described in Section B.6, models are written in the input language supported by the NuSMV model checker, and the FSAP platform provides users with the possibility of using their preferred text editor for editing the models. In the ESACS methodology, two scenarios are possible. In the first scenario, a formal model, called *Formal Safety Model* (FoSAM), is written by the safety engineer, with the intent of highlighting the main safety features of the system under modeling and assessing its architecture. In the second scenario, as described in Section 5.8, a formal model describing the nominal behavior of the system, called the *system model*, is written by the design engineer and then automatically extended with the definition of the failure modes, originating the so-called *extended system model*, which describes the behavior of the system in the presence of faults.

■ **FM Capturing.** The Failure Mode Capturing component is responsible for the definition of the failure modes to be injected into the system and is used in the second modeling scenario previously described. This block is composed of a Failure Mode Editor, for editing the definitions of the failure modes and associating them to components in the system model, and a so-called *Generic Failure Mode Library* (GFML), which contains the definitions of some predefined failure modes. FSAP supports different kinds of failure modes, namely *inverted, stuck_at ⟨value⟩, cstuck_at ⟨value⟩, random* (nondeterministic output), and *ramp_down* (compare Section 5.8.2). For each failure mode, the following characteristic can be defined: system variable the failure mode must be associated to, name, description, type, optional parameters, probability of occurrence, and activation mode (*permanent* or *sporadic*). The selected failure modes are automatically injected into the system model through a model extension facility, as discussed in Section 5.8. The FSAP platform allows the user to group different failure modes into a so-called *failure set*. A failure set can be used to group different failures that are not independent due to, for example, a common cause or a particular risk (compare Section 4.9.5). Depending on the activation model, the failure modes within a failure set can be defined either as *simultaneous* or *cascading*. In the latter case, the user can enter constraints about the relative time of activation of the different failure modes. The model extension phase also inserts additional variables in the model that allow the user to constrain, in a finer-grained fashion, the number of (simultaneous) failures that may occur.

■ **SR Capturing.** The Safety Requirements Capturing block is responsible for managing the input of safety requirements. Safety requirements can be defined either by the design or by the safety engineer, and they will be used at a later stage to assess the functional and safety behavior of the system, both in the nominal case, and in the presence of the faults. The Safety Requirements Capturing block is composed of a Safety Requirement Editor, and a so-called *Generic Safety Requirements Library* (GSRL) that contains the definitions of generic property patterns that can be used to instantiate specific properties. The GSRL comprises the most frequently used patterns that can be instantiated to yield different forms of safety, liveness, and reachability properties. Safety requirements are written in temporal logic (compare Section 5.6.6), and they can be either edited directly by the user or instantiated from the patterns in the GSRL. For each safety requirement, the following characteristic can be defined: name, description, type, and severity classification.

■ **Model Analysis.** The Model Analysis block provides the core functionalities for verification and safety assessment of a system model against the functional and safety requirements previously defined. This block is built upon the underlying formal verification tools, in particular the NuSMV-SA block, which is built upon the NuSMV model checker. The functionalities provided by FSAP include, among others, model simulation, property verification, generation of fault trees, and Failure Mode and Effects Analysis tables. The complete set of functionalities is described in more detail in Section B.7. The Model Analysis block contains the Analysis Task Manager for dealing with analysis tasks. Each analysis task is defined through an interface that allows the definition of the following features: name, description, result directory of the task, verification engine to be used and selection of the related options, and verification hypotheses. Analysis tasks are managed via the central SAT Repository, and, once they have been run, they contain links to the corresponding analysis results. The Model Analysis block also contains the Extended System Model (ESM) generator, which is responsible for calling the model extension facility whenever needed.

■ **Safety Results Extraction.** The Safety Results Extraction block processes the results produced during model analysis, and presents them in a human-readable form. The results associated with a safety analysis task can be accessed via a so-called Result Displayer interface. The following results can be displayed: (counterexample or simulation) traces, which can be viewed either within a text editor or a spreadsheet, or plotted using a graphical viewer; fault trees, which can be displayed with a graphical fault tree displayer; FMEA tables, which can be viewed either within a text editor or a spreadsheet; and ordering information (see Section B.7), which is displayed in textual form. The following information is recorded, and then shown in the Result displayer window, whenever an analysis task has been run: name of the task, time of execution, successful completion of the task, and task-related information (e.g., property validity for property verification, probability of the top-level event for Fault Tree Analysis). Finally, a link to the low-level output of the model checker is provided for debugging purposes.

■ **Safety Analysis Tools.** The Safety Analysis Tools block comprises the different tools used to display and post-process the results of model analysis (e.g., the Fault Tree Displayer and the Simulation Displayer).

B.4 Implementation

The FSAP platform is implemented partly in ANSI C (for the NuSMV-SA add-on) and in C++ (for the graphical user interface), and it can be compiled for several target operating systems. The graphical user interface is based on the FLTK (FLTK, 2009) cross-platform toolkit. All the data produced by the platform are stored in XML format, and the corresponding parser is based on the cross-platform Expat (Expat, 2009) library.

The NuSMV-SA add-on is built upon NuSMV and hence inherits the core features of NuSMV. In particular, it exploits the CUDD library (CUDD, 2009) for BDD-based model checking, and it supports the external solvers for SAT-based Bounded Model Checking that are available in NuSMV, specifically zChaff (zChaff, 2009) and MiniSat (MiniSat, 2009).

The NuSMV-SA add-on extends NuSMV with some additional commands that are specific to safety analysis. The set of commands and their interfaces are documented in the user manual of the platform, which can be downloaded from FSAP (2009).

B.5 User Interaction

FSAP has a graphical user interface that provides access to all the functionalities supported by the platform: project management, editing of models, failure modes, safety requirements, definition of analysis tasks, verification, and display of results. The platform also provides online help, logging facilities, and setting of user options and preferences. The FSAP user manual can be downloaded from the platform web page (FSAP, 2009). Screenshots of the graphical user interface can be found both in the user manual and in Bozzano and Villafiorita (2007).

The graphical user interface has been designed as the default way for user interaction. In addition, FSAP is also provided with a command-line interface that can be used both in interactive and in batch mode. The command-line interface only supports a subset of the platform functionalities; in particular, it is possible to load an existing project, generate the corresponding extended system model, and run (some or all of) the verification tasks associated with it. The command-line interface can be used, in batch mode, for debugging and benchmarking purposes.

B.6 Language

FSAP is based on the textual language supported by NuSMV (see Section A.6). In particular, FSAP supports the same data types, the modeling constructs, the composition rules, and the hierarchical modular structure of NuSMV specifications. Definition of failure modes can be associated with specific variables within a module, and therefore they are instantiated as many times as the corresponding module. As discussed in Section 5.8.2, the result of model extension (i.e., injection of the definition of the failure modes into a system model) is an automatically generated NuSMV model. Finally, formal requirements in FSAP can be written using either CTL and LTL temporal logics (see Section 5.6.6).

A collection of examples is provided with the FSAP distribution, and is also available for download from the platform web page (FSAP, 2009).

B.7 Functionalities

The FSAP platform is based on the ESACS methodology (Bozzano et al., 2003b; Åkerlund et al., 2006; Bozzano and Villafiorita, 2007), which is described in more detail in Section 5.8. Central to the methodology are the concepts of fault injection and model extension, described in Section 5.8.1 and 5.8.2, which enable model verification and safety assessment both under nominal conditions and in the presence of faults.

The main verification functionalities supported by FSAP are listed and briefly discussed as follows:

- **Model Simulation.** FSAP allows the user to perform (possibly constrained) random simulation. It is possible to simulate both the nominal and the extended system model, for a given number of steps. Model simulation is provided by standard NuSMV commands, and it generates a trace, in NuSMV format, that can be visualized within a text editor, imported into a spreadsheet, or plotted with a simulation plotter. It is possible to plot only selected variables, and to select a set of default variables to be plotted, which can be associated with a model and retrieved across different sessions.

- **Property Verification.** FSAP supports property verification capabilities that are based on standard model checking functionalities of NuSMV. A property can be verified against the nominal or the extended system model. Both BDD- and SAT-based verification can be used, and a number of possible options for the verification engines can be set. Additional hypotheses (in the form of invariants) can be defined and associated with a given verification task, in order to constrain the analysis further. The result of property verification is a yes/no result, depending on whether or not the property is verified. In case the property is not verified, a counterexample trace is generated, which can be visualized in the same way as a simulation trace.

- **Fault Tree Analysis.** The FSAP platform implements ad-hoc algorithms for Fault Tree Analysis, both in the case of *monotonic* and *non-monotonic analysis* (i.e., when the presence as well as the absence of faults is taken into account during the analysis). Fault Tree Analysis in FSAP is based on the generation of *minimal cut sets* (MCSs) (*prime implicants* in the case of non-monotonic analysis), and is implemented as additional commands in the NuSMV-SA add-on. The implementation is built on top of some routines for minimization of Boolean functions (see Coudert and Madre, 1992, 1993; Rauzy, 1993; Rauzy and Dutuit, 1997). More details about the implementation can be found in Bozzano et al. (2007) and in Section 5.8.4. The result of a Fault Tree Analysis task is a fault tree with a two-layer logical structure, consisting of the top-level disjunction of the minimal cut sets, each cut set being the conjunction of the corresponding basic faults. For each cut set, FSAP also generates an associated trace that witnesses the fact that a specific combination of faults can result in the violation of the top-level event. Additionally, each event and gate in the fault tree can be optionally labeled with a probability, that is computed, under the hypothesis of fault independence, on the basis of user-defined probabilities for the basic faults. Fault trees can be visualized with an ad-hoc graphical fault tree displayer. Moreover, they can be exported with a format that is compatible with the FaultTree$+^{©}$ format (FT+, 2009) or, alternatively, in SVG (*Scalable Vector Graphics*) format.

■ **Failure Mode and Effect Analysis.** It is possible to use FSAP to perform Failure Mode and Effects Analysis, which is implemented as an additional command in the NuSMV-SA add-on. The result of the analysis is a table that relates a set of fault configurations with a set of properties, and that can be displayed within a text editor or using a spreadsheet.

■ **Common Cause Analysis.** The FSAP platform provides facilities that ease the modeling and verification of particular risks or common causes. Specifically, it is possible to define groups of failures that are intended to occur simultaneously (or in a cascading manner, as described in Section B.3) as a result of a common initiating event. Such groups of failures are called *failure sets*. The definition of failure sets affects the way FSAP carries out some of the analyses. In particular, in Fault Tree Analysis and Failure Mode and Effects Analysis, a failure set is considered a single fault (e.g., a minimal cut set consisting of a failure set is considered a single point of failure). Given that failure sets break the hypothesis of fault independence, we must attach a basic probability to every failure set, for quantitative evaluation.

■ **Fault Detection, Isolation, and Recovery.** FSAP supports some analysis tasks related to Fault Detection, Isolation, and Recovery (FDIR). These tasks are based on the definition of (classes of) messages that represent observable system behavior. Fault recovery is intended to model the capability of a system to recover from a fault; it can be formalized with standard safety requirements and verified using property verification (where the property formalizes the notion of recovery). Fault detection and fault isolation are implemented as two additional commands in the NuSMV-SA add-on. The goal of fault detection is to identify, for any given input fault, the set of possible *detection means*— that is, the set of messages that will eventually be raised (in every possible scenario) as a consequence of the fault. This type of the analysis is useful for verifying whether it is possible to detect faults. The goal of fault isolation is to identify, for any given message, the possible faults that may have caused the message to be raised. This sort of analysis is useful in evaluating if faults can be isolated, that is, messages can be univocally associated with faults (in the scenario of perfect isolation, a message is associated with exactly one fault). The implementation of fault detection relies on model checking routines, whereas fault isolation is based on Fault Tree Analysis. The results of fault detection analysis (the detection means) can be visualized in textual form, whereas fault isolation analysis generates fault trees as output (one fault tree for each input message).

■ **Ordering Analysis.** FSAP provides an additional analysis capability to assess the results of Fault Tree Analysis. In particular, for each of the minimal cut sets of a given fault tree, it is possible to analyze the order in which the constituent basic events can occur, in order for the top-level event to show up. Ordering constraints can be due, for example, to a causality relation or a functional dependency between events, or by more complex interactions involving the dynamics of a system. The result of ordering analysis can be represented by a set of precedence graphs showing the dependencies between failure events (e.g., an event may occur before, after, or simultaneously with another one). Each precedence graph shows one possible alternative (notice that if basic events can occur in any order, then the set of precedence graphs is empty). The output of ordering analysis can be displayed in textual form. Ordering analysis can be coupled with Fault Tree Analysis, in the sense that each cut set within a fault tree (with more than one basic fault) can be analyzed in this way. More details about ordering analysis and its implementation can be found in Bozzano and Villafiorita (2003).

No constraint is put on the way in which the functionalities supported by FSAP can be invoked (short, of course, of the trivial constraints imposed by the information flow, e.g., a system model must be defined in order to be verified). The hypothesis underlying this approach is that, by selecting or combining the above-mentioned functionalities in different ways, FSAP can be used in different phases of the process and integrated in different design processes. Therefore, it is the responsibility of the user to ensure that the use of the platform is compliant with the procedures and the processes followed for system development within a specific organization.

B.8 Applications

The FSAP platform has been developed within the ESACS (Enhanced Safety Assessment for Complex Systems) (ESACS, 2009; Bozzano et al., 2003b) and ISAAC (Improvement of Safety Activities on Aeronautical Complex systems) (ISAAC, 2009; Åkerlund et al., 2006) projects, two European-Union-sponsored projects involving various research centers and industries in the fields of avionics and aerospace. FSAP is based on the methodology, called the ESACS methodology, developed within these projects. Within ISAAC, additional topics, such as a formal approach to Common Cause Analysis and to system diagnosability and testability, have been investigated. An application of the FSAP platform in an industrial setting is described in Bozzano et al. (2003a). At present, this line of work is being pursued in a follow-up project, called MISSA (More Integrated Systems Safety Assessment) (MISSA, 2009), involving most of the partners of the previous project and some additional ones. The main focus of MISSA is the application of model-based safety analysis in avionics, throughout the different phases of aircraft design and safety assessment, and the automatic construction of formal arguments supporting system safety assessment and product certification.

The FSAP platform is being used for educational and teaching purposes (it has been licensed to some academic institutions outside Fondazione Bruno Kessler), and has received interest from some industrial companies. Within the previously mentioned projects, FSAP has been used for technology transfer to the industrial partners.

The FSAP platform has also been applied in the aerospace domain. In particular, FSAP has been used within the OMC-ARE project (On-board Model Checking Autonomous Reasoning Engine) (OMC-ARE, 2009), which investigated model-based autonomous reasoning for aerospace applications (Bozzano et al., 2008a, b) (the project was evaluated in a case study involving the management of a planetary rover and an orbiting spacecraft). Within this project, FSAP provided the required functionalities for implementing fault detection and fault isolation management. More recently, FSAP has been used in the COMPASS project (COrrectness, Modeling and Performance of AeroSpace Systems) (COMPASS, 2009), which focused on the development of a toolset, called the COMPASS toolset, for the design and verification of complex applications in the aerospace domain (Bozzano et al. 2008a, b, 2009). The toolset provides capabilities ranging from formal verification to formal safety analysis and performability evaluation. The FSAP platform provides the safety analysis capabilities of the toolset, and it has been integrated with the MRMC (MRMC, 2009) probabilistic model checker to deal with performability and probabilistic analysis. Both the OMC-ARE and the COMPASS projects have been funded by the European Space Agency.

References

Åkerlund, O., P. Bieber, E. Böede et al. (2006). ISAAC, a framework for integrated safety analysis of functional, geometrical and human aspects. In *Proc. European Congress on Embedded Real Time Software (ERTS 2006)*.

Bozzano, M., G. Burte, A. Cimatti et al. (2008a). System and Software Co-Engineering: Performance and Verification. *ESA Workshop on Avionics Data, Control and Software Systems (ADCSS)*.

Bozzano, M., A. Cavallo, M. Cifaldi, L. Valacca, and A. Villafiorita (2003a). Improving safety assessment of complex systems: An industrial case study. In K. Araki, S. Gnesi, and D. Mandrioli (Eds.), *Proc. Formal Methods, International Symposium of Formal Methods Europe (FME 2003)*, Volume 2805 of *LNCS*, pp. 208–222. Berlin: Springer.

Bozzano, M., A. Cimatti, A. Guiotto et al. (2008b). On-Board Autonomy via Symbolic Model Based Reasoning. *10th ESA Workshop on Advanced Space Technologies for Robotics and Automation (ASTRA 2008)*.

Bozzano, M., A. Cimatti, J.-P. Katoen, V.Y. Nguyen, T. Noll, and M. Roveri (2009). The COMPASS approach: Correctness, modelling and performability of aerospace systems. In B. Buth, G. Rabe, and T. Seyfarth (Eds.), *Proc. 28th International Conference on Computer Safety, Reliability and Security (SAFECOMP 2009)*, Volume 5775 of *LNCS*, pp. 173–186. Berlin: Springer.

Bozzano, M., A. Cimatti, and F. Tapparo (2007). Symbolic fault tree analysis for reactive systems. In *Proc. 5th International Symposium on Automated Technology for Verification and Analysis (ATVA 2007)*, Volume 4762 of *LNCS*, pp. 162–176. Berlin: Springer.

Bozzano, M. and A. Villafiorita (2003). Integrating fault tree analysis with event ordering information. In *Proc. European Safety and Reliability Conference (ESREL 2003)*, pp. 247–254. Leiden, The Netherlands: Balkema Publisher.

Bozzano, M. and A. Villafiorita (2007). The FSAP/NuSMV-SA safety analysis platform. *Software Tools for Technology Transfer* 9(1), 5–24.

Bozzano, M., A. Villafiorita, O. Åkerlund et al. (2003b). ESACS: An integrated methodology for design and safety analysis of complex systems. In *Proc. European Safety and Reliability Conference (ESREL 2003)*, pp. 237–245. Leiden, The Netherlands: Balkema Publisher.

COMPASS (Last retrieved on November 15, 2009). The COMPASS Project. `http://compass.informatik.rwth-aachen.de`.

Coudert, O. and J.C. Madre (1992). Implicit and incremental computation of primes and essential primes of Boolean functions. In *Proc. 29th Design Automation Conference (DAC'92)*, pp. 36–39. IEEE Computer Society.

Coudert, O. and J.C. Madre (1993). Fault tree analysis: 10^{20} prime implicants and beyond. In *Proc. Annual Reliability and Maintainability Symposium (RAMS'93)*, pp. 240–245. Washington, D.C.: IEEE Computer Society.

CUDD (Last retrieved on November 15, 2009). CUDD: CU Decision Diagram Package. `http://vlsi.colorado.edu/~fabio/CUDD`.

ESACS (Last retrieved on November 15, 2009). The ESACS Project. `http://www.esacs.org`.

Expat (Last retrieved on November 15, 2009). The Expat XML Parser. `http://expat.sourceforge.net`.

FBK (Last retrieved on November 15, 2009). Fondazione Bruno Kessler. `http://www.fbk.eu`.

FLTK (Last retrieved on November 15, 2009). FLTK: Fast Light Toolkit. `http://www.fltk.org`.

FSAP (Last retrieved on November 15, 2009). The FSAP/NuSMV-SA Platform. `https://es.fbk.eu/tools/FSAP`.

FSAP2 (Last retrieved on November 15, 2009). The FSAP Safety Analysis Platform: Commercial Pages. `http://fsap.fbk.eu`.

FT+ (Last retrieved on November 15, 2009). FaultTree+. `http://www.isograph-software.com/ftpover.htm`.

ISAAC (Last retrieved on November 15, 2009). The ISAAC Project. `http://www.cert.fr/isaac`.

MiniSat (Last retrieved on November 15, 2009). The MiniSat Page. `http://minisat.se`.

MISSA (Last retrieved on November 15, 2009). The MISSA Project. `http://www.missa-fp7.eu`.

MRMC (Last retrieved on November 15, 2009). MRMC: Markov Reward Model Checker. `http://www.mrmc-tool.org`.

OMC-ARE (Last retrieved on November 15, 2009). The OMC-ARE Project. `http://es.fbk.eu/projects/esa_omc-are`.

Rauzy, A. (1993). New algorithms for fault trees analysis. *Reliability Engineering and System Safety* 40(3), 203–211.

Rauzy, A. and Y. Dutuit (1997). Exact and truncated computations of prime implicants of coherent and non-coherent fault trees within Aralia. *Reliability Engineering and System Safety* 58(2), 127–144.

zChaff (Last retrieved on November 15, 2009). zChaff. `http://www.princeton.edu/~chaff/zchaff.html`.

Appendix C

Some Regulatory Bodies and Regulations

In this appendix we briefly list and present some of the main organizations responsible for the regulation, certification, and enforcement of safety standards in various industrial sectors. Providing a directory of all the standardization bodies is a daunting task. Several agencies are established at the national level. Different industry sectors have different regulatory bodies. In some cases, industrial groups found nonprofit organizations to promote standardizations in specific industrial sectors. On top of these, we need to mention international organizations dedicated to specific sectors and various standardization bodies that publish standards in different areas.

The list presented in this appendix is therefore incomplete. It is, however, a good starting point for obtaining more information about standards, standardization, and about the certification of safety-critical systems. Moreover, many of the organizations listed below provide access to their document over the Internet, some for free and others at a cost.

C.1 BSI, British Standards Institution

Founded in 1901, BSI was the world's first standardization body and it is the United Kingdom's National Standards Body.

BSI develops and promotes standards covering the most diverse areas, from Aerospace to Food & Drink. According to its home page, BSI has about 27,000 standards (as of November 2009). Among them, we mention BS 61508 for the railway sector, also endorsed by the IEC (International Electrotechnical Commission) and mentioned in Chapter 6.

The home page of BSI is `http://www.bsigroup.com`. (This and all the other URLs presented in this appendix were last retrieved in March 2010.)

C.2 CENELEC, European Committee for Electrotechnical Standardization

CENELEC, the European Committee for Electrotechnical Standardization, is a nonprofit technical organization founded in 1973 from the merging of two European organizations, CENELCOM and CENEL. To understand the mission of CENELEC, it is worthwhile to briefly discuss its predecessors, CENELCOM and CENEL. In particular, CENELCOM and CENEL were founded to promote standardization and break trade barriers in the European Union. The IEC National Electrotechnical Committees of five European Nations participated in the creation of CENELCOM. CENEL extended CENELCOM activities by including the IEC Committees of various non-EU countries.

The main activity of CENELEC is the preparation of voluntary standards, with priority given to the standards that determine the safety and free movement of goods and services in the European Union. Of interest to the readers of this book might be, in particular, CENELEC's standards related to the railway sector. Finally, we note that, as stated by CENELEC's mission, the standardization body supports the IEC in achieving its mission.

The home page of CENELEC is `http://www.cenelec.eu`.

C.3 DIN, Deutsches Institut für Normung e. V.

DIN, the German Institute for Standardization, is a nonprofit organization that develops norms and standards for industry and governments. Founded in 1917, DIN is the acknowledged national standards body that represents German interests in European and international standards organizations.

DIN has published 33,800 standards (as of November 2009) covering the most diverse areas. The home page of DIN is `http://www.din.de`.

C.4 DSTAN, U.K. Defence Standardization

DSTAN is part of the Procurement Development Group and is the authority of the U.K. Ministry of Defense responsible for standardization. The authority plans the standardization needs for the U.K. armed forces, and it manages the adoption of civil standards and production of new standards when civil standards are not available.

As of December 2009, the agency had approximately 2000 standards, most of which are available for download through the Internet. The home page of DSTAN is `http://www.dstan.mod.uk`.

C.5 EASA, European Aviation Safety Agency

The European Aviation Safety Agency (EASA) is an agency of the European Union established in 2002. Similar to the FAA, the role of EASA is to promote safety standards in civil aviation, and the EASA is considered the centrepiece of the European Union's strategy for aviation safety.

The agency's responsibilities include

- Providing advice for legislation.
- Implementing and monitoring safety rules.
- Type-certification of aircraft and components, as well as the approval of organizations involved in the design, manufacture, and maintenance of aeronautical products. Type-certification is conducted through the European national aviation authorities.
- Authorization of third-country (non EU) operators.

The home page of EASA is `http://www.easa.eu.int`.

C.6 EUROCAE

EUROCAE, the European Organization for Civil Aviation Equipment, is a nonprofit organization established in 1963 to provide a European forum for resolving technical problems with electronic equipment for air transportation.

EUROCAE develops standards by establishing working groups (WGs) composed of members voluntarily provided by associates. EUROCAE standards go under the name of "ED" (EUROCAE Document). We mention ED-12B, which is the European equivalent of DO-178B.

The home page of EUROCAE is `http://www.eurocae.net`.

C.7 EUROCONTROL

The European Organisation for the Safety of Air Navigation (EUROCONTROL) was founded in 1960 as a civil-military organization for the management of air traffic control in the EU member states. Its mission is to sustain the growth of a pan-European air traffic management system. Among the responsibilities we note the development and uniform implementation of harmonized safety regulatory objectives and requirements for the European Air Traffic Management (ATM) system.

Starting in 2006, EUROCONTROL has led a program to improve the safety maturity of European Traffic Controls. The program includes supporting the implementation of regulations, improving incident reporting and data sharing, enhancing safety management, and improving risk assessment and mitigation.

The home page of EUROCONTROL is `http://www.eurocontrol.int`.

C.8 ERA, European Railway Agency

The European Railway Agency was set up under EC Regulation No. 881/2004 to help create an integrated railway area by reinforcing safety and interoperability. Concerning safety, the Agency issues European-wide recommendations related to the harmonization of safety assessment procedures and certification of railway systems. The Agency also develops safety indicators, defines safety targets, and monitors their achievement by the EU member states. Finally, the ERA promotes the standardization of accident investigation and reporting in the railway sector.

Concerning interoperability, the Agency is responsible of the European Rail Traffic Management System, a pan-European standard for train management.

The home page of ERA is `http://www.era.europa.eu`.

C.9 ESA, European Space Agency

The European Space Agency is an inter-governmental organization that focuses on the space exploration efforts of the European national agencies. The Agency promotes, through the European Cooperation for Space Standardization (ECSS), a set of standards for use in all European space activities. The standards cover all aspects of system development, from management to operations.

The home page of ESA is `http://www.ecss.nl/`.

C.10 FAA, Federal Aviation Administration

The Federal Aviation Administration (FAA) is an agency of the U.S. Department of Transportation and is responsible for regulating and overseeing all aspects of civil aviation in the United States. In the words of the FAA, its mission is "to provide the safest, most efficient aerospace system in the world" (FAA, 2009a).

The Agency was established in 1958, following a series of midair collisions and with the aim of properly governing the approaching introduction of jet airliners (FAA, 2009b). Among the activities of the FAA we mention

- Issuing and enforcing safety regulations and minimums; standards for development, operation, and maintenance of aircraft
- Type-certification of avionic systems
- Airspace and air traffic management, through the development of air traffic rules, by assigning the use of airspace, and by controlling air traffic

The home page of FAA is `http://www.faa.gov/`.

C.11 IEC, International Electrotechnical Commission

Founded in 1906, the IEC had Lord Kelvin as its first president. The IEC prepares and publishes international standards for electrical and electronic technologies. The standards published by IEC serve as a basis for national standards and for tenders. Being international in nature, the IEC also serves to break national barriers and thus promote international trade.

The home page of IEC is `http://www.iec.ch`.

C.12 ICAO, International Civil Aviation Organization

The International Civil Aviation Organization (ICAO) is an agency of the United Nations founded in 1944 by 52 international states that signed the "Convention on International Civil Aviation" (also known as the "Chicago Convention"). The ICAO promotes safety in the civil aviation sector by setting strategic goals, by endorsing standards, and by maintaining a set of standardized procedures for accident investigation.

The Organization consists of an Assembly, a Council, and a Secretariat. The Assembly, composed of representatives from all Contracting States, is the sovereign body of the ICAO. It meets every 3 years and establishes the policies for upcoming years. The Council is responsible for the development and adoption of standards.

The home page of ICAO is `http://www.icao.int`.

C.13 ISO, International Organization for Standardization

Founded in 1946, the ISO was born from the union of two organizations: the ISA (International Federation of the National Standardizing Associations) and the UNSCC (United Nations Standards Coordinating Committee).

The ISO is the world's largest developer and publisher of international standards. It functions as a network of the national standards institutes of 162 countries, one member per country, with a Central Secretariat in Switzerland that coordinates the system. The ISO has published more than 17,000 standards covering all industry sectors, with the exceptions of electrotechnology (covered by the IEC) and telecommunications (covered by the ITU).

Among the standards of interest for this book, we mention the process-and quality-related standards, ISO 12207 (published by the IEC under the same number), and the special adoption of some widely adopted aerospace standards. In particular, the ISO recognizes aerospace standards meeting the following criteria (ISO, 2009):

- Standard is currently in use in multiple aerospace programs or in multiple countries.
- Standard is suitable for new/future designs.
- Standard does not duplicate another standard in current international use or conflict with an existing ISO standard.
- Standard contains no requirements that would result in a barrier to trade or to use by any nation.
- Standard is available at least in English or French.
- Originating organization agrees with its submission for ISO recognition.

Among the standards recognized, we cite DO-178B (ED-12B).

The home page of ISO is `http://www.iso.org`.

C.14 JAA, Joint Aviation Authorities

The JAA were an associated body of the European Civil Aviation Conference (ECAC) representing the civil aviation regulatory authorities of a number of European states that had agreed to cooperate in developing and implementing common safety standards. The JAA were responsible for the production and publication of Joint Aviation Requirements (JARs) and the associated guidance and administrative documents.

Starting in 2003, the regulations were transposed and converted into EASA regulatory measures. The JAA role has now been taken by EASA. See the entry for EASA (Section C.5) for more details.

The home page of JAA is `http://www.jaa.nl`.

C.15 NASA, National Aeronautics and Space Administration

NASA is an agency of the U.S. Government responsible for the nation's public space program. Established in 1958 by President Dwight D. Eisenhower, NASA grew out of the National Advisory Committee on Aeronautics (NACA), which had been researching flight technology for more than 40 years. NASA has been a publisher and promoter of standards in the public and private sectors alike.

Standardization is the responsibility of the NASA Technical Standards Program (NTSP). The NTSP is responsible for

- Integrating with national and international standardization activities by promoting development of non-government standards and by providing NASA-wide access to international standards
- Developing NASA technical standards where needed

The home page of NASA is `http://www.nasa.gov`. NASA technical standards and documentation can be retrieved from the following web sites:

- The NODIS Library (NASA Online Directives Information System Library), available at `http://nodis3.gsfc.nasa.gov/`
- START (Standards and Technical Assistance Resource Tool), available at `http://standards.nasa.gov/public`
- NTRS (NASA Technical Report Server), available at `http://ntrs.nasa.gov/search.jsp`
- NASA Headquarters Library, from which we list the home page, `http://www.hq.nasa.gov/office/hqlibrary`, and `http://www.hq.nasa.gov/office/hqlibrary/find/documents.htm`, which points to other NASA resources

C.16 NRC, U.S. Nuclear Regulatory Commission

The NRC is a Commission of the U.S. Government founded in 1974, when the U.S. Government reorganized the energy sector, by assigning to the Department of Energy the production and promotion of nuclear power and to the NRC the regulatory work.

The Commission is headed by five members, who are appointed by the President and confirmed by the Senate for a 5-year period. The Commission as a whole is responsible for all policies and regulations related to the civil usage of nuclear power. Executive Directors for Operations (EDOs) are responsible for carrying out the policies and decisions of the Commission. Finally, offices reporting to the EDO ensure that the commercial use of nuclear materials in the United States is safely conducted.

Following the Three Mile Island accident, the NRC reexamined the adequacy of its safety requirements and imposed new regulations. Among the actions, the NRC put greater emphasis on the human factor and new regulations were issued concerning training, overtime, and testing and licensing.

NRC publications, NUREGs, include one of the historical references to fault trees, the *Fault Tree Handbook* (Vesely et al., 1981).

The home page of NUREG is `http://www.nrc.gov/`.

C.17 RTCA, Radio Technical Commission for Aeronautics

Founded in 1935 as the Radio Technical Commission for Aeronautics, RTCA is a private, nonprofit organization founded to advance the art and science of aviation and aviation electronic systems. RTCA includes more than two hundred government, industry and academic organizations from the Unites States and more than one hundred affiliates from around the world.

The RTCA develops standards related to the FAA. This is typically done by establishing, upon specific request, a special committee to recommend Minimum Operational Performance Standards (MOPS) or appropriate technical guidance documents. MOPS are developed by the RTCA and become the basis for certification.

The RTCA publishes standards under the "DO" name—see, for example, DO-178B.

The home page of RTCA is `http://www.rtca.org`.

C.18 RSSB, Rail Safety and Standards Board

The Rail Safety and Standards Board is a private, nonprofit organization established in 2003 to promote safety in the railway sector in Great Britain. The RSSB mission includes

- Management of railway group standards on behalf of the industry
- Development of a long-term safety strategy for the industry, including the publication of annual Railway Strategic Safety Plans
- Measurement and reporting on safety performances in the railway sector

The group publishes the "Yellow Book," a set of good practices for the development of railway applications.

The home page of the RSSB is `http://www.rssb.co.uk/`.

C.19 SAE International

The International Society of Automotive Engineers was founded in 1905 by industries and practitioners in the automobile sector to address the growing need for shared technical standards. SAE's core competencies include the development of standards, lifelong learning, and information sharing in the automotive and, more recently, aerospace sectors. SAE has developed more that 2,600 globally used and recognized standards for the ground vehicle industry and more than 6,700 documents for the aerospace sector, the largest such collection of consensus standards in the world.

SAE standards include the Aerospace Recommended Practices (ARPs), which we have discussed various times in this book.

The home page of SAE is `http://www.sae.org`.

C.20 U.S. Department of Defense

The U.S. Department of Defense (DoD) promotes, adopts, and defines standards to regulate the procurement and development of military equipment. In the 1950s and 1960s, the DoD has been a principal actor in the dissemination of standards, safety, and quality in the private sector.

The standardization effort of the DoD is carried out through the Defence Standardization Program. Among the goals are integration with civil standards, interoperability with civil and other Allies' standards, and the development and management of standards for the military sector.

The starting points include `http://www.dsp.dla.mil`, the home page of the Defense Standardization Program, and ASSIST-Online `https://assist.daps.dla.mil/online/start/`, which provides unified access to several standards.

References

FAA (Last retrieved on November 15, 2009a). FAA mission. Available at `http://www.faa.gov/about/mission/`.

FAA (Last retrieved on November 15, 2009b). The Federal Aviation Administration: A historical perspective, 1903–2008. Available at `http://www.faa.gov/about/history/historical_perspective`.

ISO (Last retrieved on November 15, 2009). Directory of iso-recognized international aerospace standards. Available at `http://www.iso.org/iso/iso_catalogue/directory_of_aerospace_standards.htm`.

Vesely, W.E., F.F. Goldberg, N.H. Roberts, and D.F. Haasl (1981). Fault Tree Handbook. Technical Report NUREG-0492, Systems and Reliability Research Office of Nuclear Regulatory Research U.S. Nuclear Regulatory Commission.

Index

Milton Keynes UK
Ingram Content Group UK Ltd.
UKHW031129141024
449569UK00006B/340